MATHEMATICS RESEARCH DEVELOPMENTS

THE FIBONACCI NUMBERS AND INTEGER STRUCTURE

FOUNDATIONS FOR A MODERN QUADRIVIUM

MATHEMATICS RESEARCH DEVELOPMENTS

Additional books in this series can be found on Nova's website under the Series tab.

Additional e-books in this series can be found on Nova's website under the eBooks tab.

MATHEMATICS RESEARCH DEVELOPMENTS

THE FIBONACCI NUMBERS AND INTEGER STRUCTURE

FOUNDATIONS FOR A MODERN QUADRIVIUM

ANTHONY G. SHANNON

AND

JEAN V. LEYENDEKKERS

nova
science publishers
New York

Library of Congress Cataloging-in-Publication Data

ISBN: 978-1-53613-454-4

Published by Nova Science Publishers, Inc. † New York

CONTENTS

FOREWORD

Krassimir Atanassov

In 1984, I published a paper in the area of number theory with some open problems that I had not yet solved. Half a year later, I received an air-mail letter from Tony Shannon with a paper that contained the solution to one of these problems. (Let me remind younger readers that in that time there was no Internet nor were there e-mails!) Tony invited me to be his coauthor of this paper. I decided that it would be not appropriate for my name only to appear without contributing so I wrote a couple of additional pages. On the other hand, in my letter of reply I put the draft of another paper, inviting Tony to be my coauthor. So, during the next ten years, in air-mail letters, we prepared and published about ten joint papers.

In 1994, two conferences were organized in Bulgaria in the areas of combinatorial mathematics and of fuzzy sets theory. For both of these conferences we prepared joint communications which were accepted, and Tony visited me in Sofia the first time. For the fuzzy sets conference in particular, we prepared a paper in which for the very first time we introduced the concept of the intuitionistic fuzzy graph.

Before our meeting in person, I knew that Tony worked in the area of number theory. At the time of our initial meeting, I understood that he was

Dean of the Graduate Research School and Professor of Applied Mathematics in his university in Australia and that he also had very serious interests in the area of mathematical modeling in biology and medicine, that he was the mathematician who developed a Binet formula for the tribonacci sequence, and so on, going back many years to the 1960s. He was the first foreigner, who started to be actively interested in applying the generalized nets as mathematical tool for the modeling of parallel processes, extending Petri nets, over which I had worked already for twelve years.

During each of the next twenty years, Tony visited me at the Bulgarian Academy of Sciences in Sofia. Consequently we wrote and published more than one hundred and fifty joint papers and fourteen books in which we, with many colleagues from different parts of the world, applied generalized nets and fuzzy logic in a variety of applications in science, medicine, engineering, mathematics and education - a reasonable measure of our continuing scientific collaboration.

At the time of our first meeting, we discussed the idea given to me by Professor Aldo Peretti from Buenos Aires, Argentina, that we publish an international journal, Notes on Number Theory and Discrete Mathematics. In the next year its first volume was published and it has been continuing now for twenty three years. I am very happy that extracts from, and generalizations, extensions and developments of, some of the ideas in these papers in Notes on Number Theory and Discrete Mathematics, appear in this book, mostly published with Jean Leyendekkers. She has not been to Bulgaria, but I have learnt a lot about her from these papers and of the numerous papers she wrote in physical chemistry for which the University of Sydney awarded her the higher doctoral degree of Doctor of Science.

The present book contains many new inter-related ideas of Tony and Jean over which they have worked in education and mathematics for many years and which Tony and I have discussed during our meetings in many different parts of Bulgaria.

I hope that these ideas will be of interest to the readers of this book, and what is even more important that will stimulate the generation of new ideas for the development of parts of number theory itself and for its use in the classroom to inspire future generations. I wish the success of the book and the continuing success of its authors!

Bulgaria
18 November 2017

PREFACE

This book has been prepared for

- teachers of mathematics at all levels for enrichment ideas (*cf.* Sellars 2017),
- students of mathematics at the senior undergraduate level for project topics, and other
- lovers of mathematics who are fascinated by the ubiquitous Fibonacci numbers (*cf.* Matthys and Wilson, 2016): 'amateurs' in the true sense of the word.

The structure of the integer system is most easily delineated via modular rings. They are essentially sorting devices that group the integers which can be placed in an array where the columns form classes which are characterized by linear equations - with the rows as the variables.

For example, the modular ring Z_4 has 4 columns of integers with as many rows as we wish to consider. In this ring the odd integers have only one class with even powers while the even integers have all powers grouped into one class. This shows many of the integers in class $\bar{1}_4$ (the integers with a residue of 1 modulo 4) to be equal to a sum of an even and odd square. In turn this class provides the main component of ALL primitive Pythagorean triples. Furthermore, the odd integers in this class provide the variable a for the generalized Golden ratio family, $\varphi_a = \frac{1}{2}\left(1+\sqrt{a}\right)$, $a = 5$ for the ordinary golden ratio.

These modular rings, because they cut back the integers to their simplest forms, can identify a variety of integer characteristics, not otherwise obvious. For instance, solutions can be established for conjectures such as the Brocard-Ramanujan and Erdös-Strauss in the context of the modular ring Z_4.

Moreover, sequences of primes can be formulated using the simple structural concept of right-end-digits (REDs or residues modulo 10, also known as unit-digits). In particular, REDs can be used to demonstrate restrictions on powers. In this monograph we apply this analysis of integer structure to a variety of number theoretic questions, especially the infinite sequence of the Fibonacci numbers and various generalizations of the associated recurrence relation or the initial conditions.

At different stages we consider aspects of Diophantine equations, particularly in Chapter 2. Diophantine equations are sometimes referred to as 'taxicab' equations in reference to the well-known story (in the film "the Man Who Knew Infinity" which was based on the book by Robert Kanigel (1991)) of how Godfrey Harold Hardy, on visiting the very sick Srinivasa Ramanujan, said that the number, 1729, of the taxicab he had ridden in was not especially interesting (Hardy 1937). Ramanjuan immediately pointed out that it is the smallest natural number which can be written in two different ways, namely,

$$1729 = 10^3 + 9^3 = 12^3 + 1^3.$$

An Indian postgraduate student has used the four digits 1, 7, 2, 9 in that order of 1729, now known as Ramanujan's number (or the Hardy-Ramanujan number), to obtain interesting expressions for all integers in the range [-529, + 529] (Sagar 2017). The number itself was known as early as 1657 (Berndt and Bhargava 1993). The smallest number expressible in n ways as a sum of cubes is called the nth taxicab number (Guy 1994). The first two taxicab numbers are 2 and 1729. In passing, one can note that the irrational constant $R = e^{\pi\sqrt{163}}$ is known as Ramanujan's constant which is 'very close' to an integer. Such numbers can be found using modular functions (*cf.* Hermite 1859).

The original (or ordinary) Fibonacci sequence was recorded before 200BC by Pingala, an Indian Sanskrit grammarian and mathematician in his book *Chandahsutra* (*cf.* Hall 2008 who looks at the combinatorics and she also links mathematics and poetry). Leonardo of Pisa or Leonardo, son (filius) of Bonaccus (Fibonacci), was born in 1170 and published two major works, *Liber Abacci* (1202) and *Liber Quadratorum* (1228) (McClenon 1919). The former book featured many mathematical problems which he solved and he is considered the greatest medieval European mathematician. The eponymous sequence, in which each member is generated from the sum of the two previous elements, was produced by a bio-mathematical model of the breeding habits of rabbits. The initial two elements are crucial and varying them gives rise to other sequences, such as that of Lucas, which is also explored in this book. We shall also see the relation of the ratio of adjacent Fibonacci numbers to the golden ratio, well-known in classical Greek architecture and revived by some major twentieth century architects (Livio 2002).

ACKNOWLEDGMENTS

Appreciation is gratefully acknowledged of the permission from Professor Krassimir T Atanassov, Institute of Biophysics and Biomedical Engineering, Bulgarian Academy of Sciences, PO Box 12, Sofia 1113, BULGARIA.

Professor József Sándor, Department of Mathematics, Babeș-Bolyai University. 537 099 Forteni nr. 83, R-Jud. Harghita, ROMANIA,

Editors-in-Chief of the international journal Notes on Number Theory and Discrete Mathematics, an official publication of the "Marin Drinov" Academic Publishing House of the Bulgarian Academy of Sciences, to use (in whole or part) material published in that journal by the authors of this book. Any such use is referenced where it occurs directly; its indirect influence can be seen in the papers cited in the bibliography.

Gratitude is also expressed to Anna Shelmerdine, Librarian at the Australian Institute of Music in Sydney, for organising and formatting the extensive bibliographic section into a Chicago style.

LIST OF FIGURES

LIST OF TABLES

Chapter 1

INTRODUCTION

Integer Structure Analysis in number theory has commonly been applied to primes, powers, various algebraic equations, and so on. However, any system which features integers by means of integer structure can reveal simple relationships that might be obscured by more sophisticated techniques. This is illustrated here where we start to transfer converging infinite series into modular ring language. We also indicate the role of number theory as an important part of a genuine liberal education accessible to all students in a way that education in the ancient *quadrivium* was confined to a small section of society. The topics of number theory in the hands of well-educated teachers can inspire a love of learning beyond passing fads.

For this reason we have embedded relevant issues on liberal education as a foundation for education in the 21st century, particularly in fostering creativity through the inspiration and passion of teachers – something which seems to wax and wane in pedagogical fashion every half-century or so (cf. Dewey 1916; Shannon 1968; Wang & Huang 2017). These issues are teased out to some extent in the first and last chapters, with the mathematics as the meat in the sandwich! One of these topics is the golden section to which the Fibonacci numbers are related. Mathematically the golden ratio is a humble surd; by replacing its argument we are able to show that it has an infinity of close relatives which can be a source of further exploration.

The context of this book is the teaching and learning of mathematics. This happens in historical and sociological contexts, though not in the way that those who are wedded to identity politics would now have us believe (*cf.* Gutiérrez, 2018). These are only one side of the mathematics education coin: the other side is in the improvement of instructional based strategies for teaching and learning (Stephens 2011).

There are sufficient historical and philosophical allusions in the development here in this short monograph for anyone to see that the mathematics *per se* transcends race, religion, history and geography.

1.1. What the Book Is About

This book is a collection of current research ideas on classical problems linked to the sequence of Fibonacci numbers. The book goes further in that it uses 'elegance' (number theory beauty) to relate some of the topics to the wider curriculum. In fact, number theory as we now know it is, in a real sense, one part of the modern version of the ancient (mathematical based) *quadrivium*. It serves as a foundation for the modern *trivium* of history, philosophy and language, all of which are of concern to the authors in what at first glance appear to be solely mathematical problems. Some of the implicit properties which we shall consider seem to have been known in classical times in Greece (Deakin 2013) and India (Hall 2008). In that sense, some familiarity of major historical landmarks and philosophical questions of number theory is part of everyone's cultural heritage of a truly liberating education.

The unifying and underlying parallel theme is the structure of the integers stripped bare by modular rings, but with references to the relevant history, philosophy and language: *history* because mathematical insights are made at a particular time; *philosophy* because of the use of logic to provide elegant solutions; *language* so that there is precision in the elaboration of ideas. That said, the book is not essentially philosophical, but the broader context cannot be ignored (Schinzel 1990).

In a mathematical sense, the book is fundamentally elementary. A consequence of this is that the problems are generally easily stated even if their elaboration requires detailed exposition. Thus the problems are simple, but the attempted solutions are neither simple nor easy. Some though are trivial, and some are tedious, but all can provide enrichment material from high school levels (if the teachers know about them) through to early postgraduate research. The topics here range from generalizations of Pythagorean problems through continued fraction algorithms to extensions of the golden section.

The classical conjectures in number theory are, by and large, inherently easy to state and intrinsically enjoyable to solve, whether it be by "mucking around" with trial and error or by systematic use of sophisticated technical tools (Franklin 2016). For these reasons, number theory can be made attractive to a wide variety of students to inspire them with notation as a tool of thought (Iverson, 1980; Halmos 1985; Hardy 1940).

These are part of general mathematical education when it goes beyond moving symbols around in an almost blind attempt to get the answer at the back of the book! As the British mathematician Peter Larcombe (2018) has vividly expressed it: "…the emotions that bubble up from connecting with a subject whose grace and allure lies at its core and awakens the spirits. Henri Poincaré's areas of interest matched broadly those of Weyl, and the French academic thought that mathematics had a triple end, one of which was that its innate aesthetic properties. In addition to stimulating enquiry within nature and philosophy should touch practitioners in ways that painting and music do." These sentiments echo in turn the thrust of Howard Gardner (2011): "Truth and beauty are fundamentally different: whereas truth is a *property of statements*, beauty reveals itself in the course of an *experience with an object.*

Although some of this material has already been published elsewhere, the difference here is that we are shedding new light on old material by connecting items hitherto diverse and by comments on what is not otherwise obvious. In particular, the two themes of the book have previously not been inter-connected in quite this way or collated this way within each of the themes. There is also a substantial quantity of new material related to the themes of the book which has not appeared before. In fact, each section has new material.

Each of the theoretical developments is accompanied by practical examples, including frequent use of tabular demonstrations so that the reader can obtain immediate reinforcement of the fundamental ideas being elaborated and extended. The book is not intended to be a stand-alone textbook as is outlined in Section 1.2, but the examples proposed in each section lend themselves to other generalizations, extensions and connections. In the words of Phillips (2005): "In the world of music, there are *two* sets of people; active musicians who play musical instruments or sing, and the much larger set of passive musicians who listen to the sounds produced by the first set. However, in the world of mathematics, we contend that there is only *one* set of mathematicians: the active set."

Thus this book is particularly suited for those teachers with a passion for the subject; teachers who wish to enlarge the horizons of their students through enrichment material either in specific mathematics classes or more general liberal arts programs.

The foregoing does raise the question "does Mathematics have a Place in the Liberal Arts" which can only be answered after we have considered the nature and scope of the liberal arts. "Such concepts as unity, truth, beauty and causality arise naturally in mathematics and an understanding of them is necessary for mastery of philosophy and theology. Perhaps the most surprising aspect of mathematics is its beauty. It is a subject which deals with absolute truth" (Townsend, 2015).

The Liberal Arts deal with the human being as a whole and hence with what lies at the essence of being human. As a result, the Liberal Arts have a far greater capacity to do good than other fields of study, for their foundation in philosophy and theology enables them to bring students into contact with the ultimate questions which they are free to accept (or reject). Even if these questions have little or no "market value," it should be obvious that the way they are taught and learned is going to have a powerful impact upon the future of the students and society.

1.2. How the Book Flows

The book is partly whimsical in the style of the previous section in that it connects with the wider panorama of the beauty and elegance of number theory and its connections with the cultural heritage of mathematics in particular, but education in general.

It is a common mistake to blame the current materialism and moral decline of the Western world on its extraordinary technological achievements, as if a scientific-technological outlook on life were incompatible with belief in God and thc supremacy of spiritual values. The decline has occurred because of the gradual displacement and internal disorientation of the properly conceived liberal arts program which should occupy the center of secondary and university education.

The late Rev Dr Austin Woodbury *sm* provocatively considered both *liber* (free) and *libra* (balance) in the etymology of "liberal." These words can be applied to Liberal Education which should free the minds of students to be open to a balanced view of the things that matter in life so that they can make decisions with an informed conscience (Newman, 1992). Nussbaum puts it well when she says that "cultivated capacities for critical thinking and reflection are crucial in keeping democracies alive and wide awake" (Nussbaum, 2010).

According to Professor David J. Walsh of the Catholic University of America, one finds in American higher education no clear idea of the end result to be aimed at. In most universities there is "an assemblage of incoherent, fragmentary disciplines and sub-disciplines... without any clearer guidance than some vague commitment to methodological requirements within the separate fields" (Walsh, 1985). In short, there is no "unifying sense of direction." But this phenomenon is not only a crisis of educators; it is a "crisis of knowledge" (Newman, 1996). Contemporary education is hopelessly at sea because, despite often vast knowledge in particular fields, many scholars lack knowledge of what matters most of all – the purpose of human existence. Again according to Walsh, "the clearest evocation of paradigmatic excellence" has traditionally been found within the cluster of disciplines called the Liberal Arts. Today, however, "education in the Liberal Arts has sadly very little to do with the formation of existential purpose; ... it has generally devolved into an increasingly irrelevant discussion of 'ideas,' 'theories,' methods and techniques." In other words, teachers of the Liberal Arts have lost their sense of vocation. Too often, so-called liberal studies can merely be a smorgasbord of subjects from which students can choose and which they tick off as they meet degree requirements. There is little sense of their inter-relations and intimate connections.

Perhaps Professor Walsh is claiming too much for the liberal arts? What does he mean by their traditionally providing "the clearest evocation of paradigmatic excellence?"

Great literature speaks both to the heart and to the mind, as do all the arts when true to their proper nature. Great literature conveys a vision of truth and beauty and moral excellence capable of raising the spirit of the reader to unsuspected heights, even in the most unpromising circumstances (Gardner, 2011). This has been demonstrated time and time again. One recalls the dramatic effect the reading of Cicero's (now lost) work, Hortensius, had on the young Augustine, kindling in him a passion for wisdom that was to inspire his whole life. For Cicero the liberal arts formed the basis of one's "humanitas" (Augustine, 1997).

In ancient Greek and Roman cultures the liberal arts referred to intellectual arts as distinct from the mechanical arts, the arts of the hand. It may be that the Greeks, in particular, exaggerated the distinction between the two kinds of art, and that the ancients in general demeaned manual work as servile. This appeared to be a mistake to many who came later, but they should not forget that the Greek conviction that reality is intelligible made possible the modern scientific revolution of which we are the heirs (Whitehead, 1994). In ancient times the seven liberal arts were the trivium and the quadrivium united by metaphysics and theology.

Contrary to the popular impression that the arts are a "soft option," true liberal education is actually very demanding. For this reason few people have attained so noble and so realistic an understanding of human affairs as for instance, Thucydides and Aristotle. Educational reformers would do well to consider including The Peloponnesian War, The Nicomachean

Ethics, and The Politics in the reading program of students who are bent upon careers in public service which, fundamentally, is a noble vocation (Maritain and Adler, 1940).

The *trivium* consisted of • Grammar→ basic systematic knowledge • Rhetoric →the 'how' • Logic →the 'why' →	The *quadrivium* consisted of • Arithmetic→Number in itself • Geometry→ Number in space • Music→ Number in time • Astronomy → Number in space and time
Illuminated by • Metaphysics → ultimate reality • Theology→ ultimate end →	Permeated by • Truth • Beauty • Goodness

Figure 1.2.1. Trivium and Quadrivium.

In answer to the question which was posed at the beginning of this section it should be clear by now that the answer is "yes," but it depends on how mathematics is taught. Thus we get to the structure of the book which itself defies mathematical ordering, whether linear, partial or quasi (Shrőder, 2002). The following diagram (Figure 1.2.1) tries to give some coherence to the topics which actually have many inter-twined links:

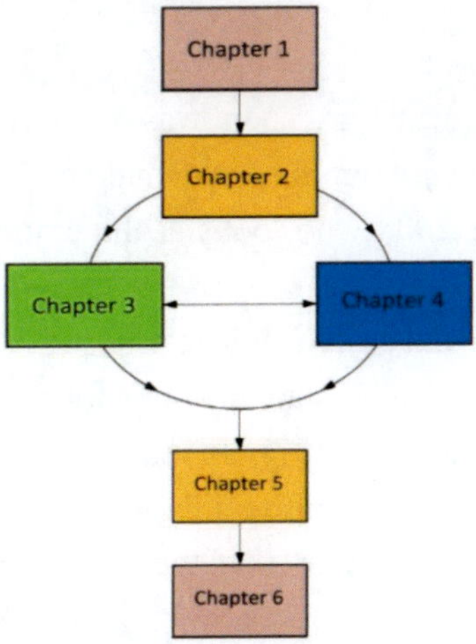

Figure 1.2.2. Flowchart of main ideas.

There is necessarily some repetition, such as repeating the Fibonacci recurrence relation [Equation 1.3.1], because we do not assume that readers will move from Chapter 1 to Chapter 6 in any particular order. Some sections will have more immediate appeal to some readers depending on their background and interests.

1.3. PRELIMINARY REMARKS[1]

As an introduction to the two major themes of this book, various Fibonacci number identities are [8] outlined here in terms of their underlying integer structure in the modular ring Z_5.

Over eight centuries ago in Chapter XII of his book, *Liber Abaci*, Leonardo of Pisa (nicknamed Fibonacci) presented and solved his famous problem on the reproduction of rabbits in terms of the famous sequence which bears his name. Four centuries later, Albert Girard (1634) gave the notation for the recurrence relation for the terms of the sequence in use today, namely

$$F_{n+1} = F_n + F_{n-1}. \tag{1.3.1}$$

The Fibonacci sequence as such was recorded before 200 BC by Pingala, an Indian Sanskrit grammarian and mathematician in his book *Chandahsutra* (*cf.* Hall 2008 who looks at the combinatorics and she also links mathematics and poetry).

Over the centuries since the Fibonacci sequence of integers has been applied to a myriad of mathematical applications, especially in number theory (Leyendekkers et al. 2007). In particular, Kepler (Livio, 2006) observed that the ratio of consecutive Fibonacci numbers converges to the Golden Ratio φ. He also showed that the square of any term differs by unity from the product of the two adjacent terms in the sequence (Simson's or Cassini's Identity (1.3.3) below).

[1] Based on Leyendekkers &.Shannon (2013a).

Table 1.3.1. Rows of modular ring Z_5

Class	$\bar{0}_5$	$\bar{1}_5$	$\bar{2}_5$	$\bar{3}_5$	$\bar{4}_5$
Row	$5r_0$	$5r_1+1$	$5r_2+2$	$5r_3+3$	$5r_4+4$
0	0	1	2	3	4
1	5	6	7	8	9
2	10	11	12	13	14
3	15	16	17	18	19
4	20	21	22	23	24

In this section we discuss the structure of the Fibonacci sequence in the context of the modular ring Z_5 (Table 1.3.1) (Leyendekkers & Shannon 2012b). The underlying structure accounts for many of the unique properties of this fascinating sequence, particularly their congruence properties (Shannon et al. 1974).

There are many characteristics of the Fibonacci sequence that are directly related to this structure. We consider some of them here. They serve as examples for further analysis. Another approach would be to consider the algebra of $F(\sqrt{5})$ where $F(x)$ is the characteristic polynomial associated with the recurrence relation (1.1) and is irreducible in the field F of its coefficients (Simons and Wright 2008).

The pattern of the Fibonacci numbers in Z_5 is displayed in Table 1.3.2.

Table 1.3.2. Fibonacci numbers in Z_5

n	1	2	3	4	5	6	7	8	9
Z_5	$\bar{1}_5$	$\bar{1}_5$	$\bar{2}_5$	$\bar{3}_5$	$\bar{0}_5$	$\bar{3}_5$	$\bar{3}_5$	$\bar{1}_5$	$\bar{4}_5$
10	11	12	13	14	15	16	17	18	19
$\bar{0}_5$	$\bar{4}_5$	$\bar{4}_5$	$\bar{3}_5$	$\bar{2}_5$	$\bar{0}_5$	$\bar{2}_5$	$\bar{2}_5$	$\bar{4}_5$	$\bar{1}_5$
n	20	21	22	23	24	25	26	27	28
Z_5	$\bar{0}_5$	$\bar{1}_5$	$\bar{1}_5$	$\bar{2}_5$	$\bar{3}_5$	$\bar{0}_5$	$\bar{3}_5$	$\bar{3}_5$	$\bar{1}_5$
29	30	31	32	33	34	35	36	37	38
$\bar{4}_5$	$\bar{0}_5$	$\bar{4}_5$	$\bar{4}_5$	$\bar{3}_5$	$\bar{2}_5$	$\bar{0}_5$	$\bar{2}_5$	$\bar{2}_5$	$\bar{4}_5$

Table 1.3.3. Details of the patterns (F_n^*: Class of F_n)

F_n^*	n for $\overline{N}_5$	n for $\overline{M}_5$
(1,6) $\bar{1}_5$	19, 39, 59, 79, ... 1, 21, 41, 61, ... 2, 22, 42, 62, ...	8, 28, 48, 68, ...
(2,7) $\bar{2}_5$	14, 34, 54, 74, ... 16, 36, 56, 76, ... 17, 37, 57, 77, ...	3, 23, 43, 63, ...
(3,8) $\bar{3}_5$	4, 24, 44, 64, ... 6, 26, 46, 66, ... 7, 27, 47, 67, ...	13, 33, 53, 73, ...
(4,9) $\bar{4}_5$	9, 29, 49, 69, ... 11, 31, 51, 71, ... 12, 32, 52, 72, ...	18, 38, 58, 78, ...
(0,5) $\bar{0}_5$	0,5,10,15,20,...	

Table 1.3.4. Data from Tables 1.3.2, 1.3.3

n	F_n^*	$\left(F_n^2\right)^*$	F_{n+1}^*	F_{n-1}^*	$\left(F_{n+1}F_{n-1}\right)^*$
4	3,8	9,4	0,5	2,7	0,5
41	1,6	1,6	1,6	0,5	0,5
77	2,7	4,9	4,9	2,7	3,8
92	4,9	6,1	3,8	4,9	7,2

The patterns of the modular residues follow the form $\overline{N}_5\bar{0}_5\overline{N}_5\overline{N}_5\overline{M}_5$ in which the numbers $\overline{N}_5$ have the pattern $\bar{1}_5\bar{3}_5\bar{4}_5\bar{2}_5$ and the interstitial numbers $\overline{M}_5$ have the pattern $\bar{2}_5\bar{1}_5\bar{3}_5\bar{4}_5$. These patterns allow prediction of the class of F_n, and hence the right-end-digit (RED) from n (Table 1.3.3).

If the measure of a line AB is given by F_{n+1},nd AB is divided into two different sized segments, AC and CB, with CB>AC, then AB/CB = CB/AC approximately defines φ, the Golden Ratio if CB = F_n and AC = F_{n-1}, so that approximately

$$F_{n+1}F_{n-1} \cong F_n^2 \tag{1.3.2}$$

However, as first noted by Kepler the two sides of (1.3.2) always differ by unity as we can see from the class structures (Table 1.3.4).

The equality is expressed in Simson's Identity

$$F_{n+1}F_{n-1} = F_n^2 + (-1)^n \tag{1.3.3}$$

$$F_{n+1} / F_n = F_n / F_{n-1} + (-1)^n / (F_n F_{n-1}) \tag{1.3.4}$$

of which the second term on the right hand side is very small for large n; this is the error term in the Fibonacci approximation for φ (Leyendekkers and Shannon 2014a).

$$F_{n+6} = 4F_{n+3} + F_n \tag{1.3.5}$$

so that

$$F_{n+6} / F_n - 4F_{n+3} / F_n = 1, \tag{1.3.6}$$

which when substituted into (1.3.3) yield

$$F_{n+1}F_nF_{n-1} = \begin{cases} F_n^3 + F_{n+6} - 4F_{n+3}, & n \text{ even}, \\ F_n^3 - F_{n+6} + 4F_{n+3}, & n \text{ odd}. \end{cases} \tag{1.3.7}$$

An odd number of golden rectangles with sides equal to successive Fibonacci numbers can appear to fit into squares as "demonstrated" in Figure 1.3.1.

This is not drawn to scale, but essentially a golden rectangle of sides F_5 and F_7 units, and hence of area 65 square units, is transformed into a square of side F_6 units and hence of area 64 square units. Of course, while the eye might just be deceived, Simson is not! From Simson's identity we get for odd n that

$$F_{n-1}^2 + F_nF_{n-1} + F_nF_{n+1} = F_{n+1}^2 \qquad (1.3.8)$$

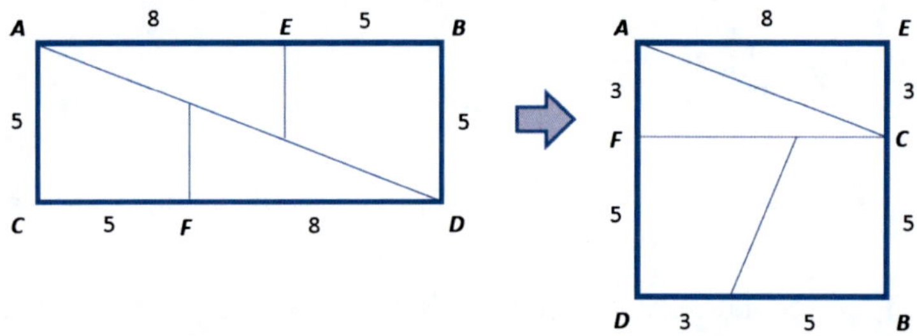

Figure 1.3.1. "Squaring" the Golden Rectangle.

Table 1.3.5. Classes of Sums

Number of Products n	Classes of		Class of
	F_{n+1}	F_{n+1}^2	$F_{n-1}^2 + F_nF_{n-1} + F_nF_{n+1}$
3	$\bar{3}_5$	$\bar{4}_5$	$\bar{1}_5 + \bar{2}_5 + \bar{1}_5 = \bar{4}_5$
5	$\bar{3}_5$	$\bar{4}_5$	$\bar{4}_5 + \bar{0}_5 + \bar{0}_5 = \bar{4}_5$
7	$\bar{1}_5$	$\bar{1}_5$	$\bar{4}_5 + \bar{4}_5 + \bar{3}_5 = \bar{1}_5$
9	$\bar{0}_5$	$\bar{0}_5$	$\bar{1}_5 + \bar{4}_5 + \bar{0}_5 = \bar{0}_5$
11	$\bar{4}_5$	$\bar{1}_5$	$\bar{0}_5 + \bar{0}_5 + \bar{1}_5 = \bar{1}_5$
13	$\bar{2}_5$	$\bar{4}_5$	$\bar{1}_5 + \bar{2}_5 + \bar{1}_5 = \bar{4}_5$
15	$\bar{2}_5$	$\bar{4}_5$	$\bar{4}_5 + \bar{0}_5 + \bar{0}_5 = \bar{4}_5$
17	$\bar{4}_5$	$\bar{1}_5$	$\bar{4}_5 + \bar{4}_5 + \bar{3}_5 = \bar{1}_5$
19	$\bar{0}_5$	$\bar{0}_5$	$\bar{1}_5 + \bar{4}_5 + \bar{0}_5 = \bar{0}_5$
21	$\bar{1}_5$	$\bar{1}_5$	$\bar{0}_5 + \bar{0}_5 + \bar{1}_5 = \bar{1}_5$

The structure of the Fibonacci sequence in Z_5 above shows that this square sum depends on the constraints of the squares which occur only in Classes $\bar{0}_5$, $\bar{1}_5$ and $\bar{4}_5$, and the sums are also confined to these classes in harmony with this square (Table 1.3.5).

The result

$$\lim_{n\to\infty} F_{n+6} / F_n = 8\varphi + 5$$

from (Leyendekkers and Shannon 2014) was obtained from the above characteristics of the sequence. Moreover, the Class of the sum of ten consecutive integers is the same as the class of the seventh number in the ten. The seventh number times 11 equals the sum of the ten. This is consistent with $11 \in \bar{1}_5$ and $\bar{1}_5 \times \bar{a}_5 = \bar{a}_5$ (Table 6). Note that the RED of the sum is the same as the RED of the seventh number, and since the RED of 11 is 1 it is the only integer to satisfy.

Table 1.3.6. Class structure in sets of 10 integers

Range of n	Class of sum	Class of 7th Integer, N_7	Class of $\bar{1}_5 \times N_7$
1 – 10	$\bar{3}_5$	$\bar{3}_5$	$\bar{3}_5$
2 – 11	$\bar{1}_5$	$\bar{1}_5$	$\bar{1}_5$
3 – 12	$\bar{4}_5$	$\bar{4}_5$	$\bar{4}_5$
4 – 13	$\bar{0}_5$	$\bar{0}_5$	$\bar{0}_5$
5 – 14	$\bar{4}_5$	$\bar{4}_5$	$\bar{4}_5$
6 – 15	$\bar{4}_5$	$\bar{4}_5$	$\bar{4}_5$
7 – 16	$\bar{3}_5$	$\bar{3}_5$	$\bar{3}_5$
8 – 17	$\bar{2}_5$	$\bar{2}_5$	$\bar{2}_5$
9 – 18	$\bar{0}_5$	$\bar{0}_5$	$\bar{0}_5$
10 – 19	$\bar{2}_5$	$\bar{2}_5$	$\bar{2}_5$

Table 1.3.7. Periodicities of Fibonacci REDs

Class of F_n	F_n^*	n^*	n	Δn
$\overline{0}_5$	0	0	30, 60, 90, 120, 150	30, 30, 30, 30
	5	5	15, 45, 75, 105, 135	30, 30, 30, 30
$\overline{1}_5$	6	1	21, 81, 141, 201	60, 60, 60, 60
		2	42, 102, 162, 222	60, 60, 60, 60
		8	48, 108, 168, 228	60, 60, 60, 60
		9	39, 99, 159, 219	60, 60, 60, 60
	1	1	1, 41, 61, 101, 121	40, 20, 40, 20
		2	2, 22, 62, 82, 122	20, 40, 20, 40
		8	8, 28, 68, 88, 128	20, 40, 20, 40
		9	19, 59, 79, 119, 139	40, 20, 40, 20
$\overline{2}_5$	2	3	3, 63, 123, 183	60, 60, 60, 60
		4	54, 114, 174, 234	60, 60, 60, 60
		6	36, 96, 156, 216	60, 60, 60, 60
		7	57, 117, 177, 237	60, 60, 60, 60
	7	3	23, 43, 83, 103, 143	20, 40, 20, 40
		4	14, 34, 74, 94, 134	20, 40, 20, 40
		6	16, 56, 76, 116, 136	40, 20, 40, 20
		7	17, 37, 77, 97, 137	20, 40, 20, 40
$\overline{3}_5$	8	3	33, 93, 153, 213	60, 60, 60, 60
		4	24, 84, 144, 204	60, 60, 60, 60
		6	6, 66, 126, 186	60, 60, 60, 60
		7	27, 87.147, 207	60, 60, 60, 60
	3	3	13, 53, 73, 113, 133	40, 20, 40, 20
		4	4, 44, 64, 104, 124	40, 20, 40, 20
		6	26, 46, 86, 106, 146	20, 40, 20, 40
		7	7, 47, 67, 107, 127	40, 20, 40, 20
$\overline{4}_5$	4	1	51, 111, 171, 231	60, 60, 60, 60
		2	12, 72, 132, 192	60, 60, 60, 60
		8	18, 78, 138, 198	60, 60, 60, 60
		9	9, 69, 129, 189	60, 60, 60, 60
	9	1	11, 31, 71, 91, 131	20, 40, 20, 40
		2	32, 52, 92, 112, 152	20, 40, 20, 40
		8	38, 58, 98, 118, 158	20, 40, 20, 40
		9	29, 49, 89, 109, 149	20, 40, 20, 40

The class structure of the 7th number in each set of ten integers is:

$$\bar{3}_5 \quad \bar{1}_5 \quad \bar{4}_5 \quad \bar{0}_5 \quad \bar{4}_5 \quad \bar{4}_5 \quad \bar{3}_5 \quad \bar{2}_5 \quad \bar{0}_5 \quad \bar{2}_5$$

which corresponds to F_7 on the F_n class pattern above.

A RED periodicity of 60 for integers was discovered in general in 1774 by Joseph Louis Lagrange (Livio 2002). However, this periodicity pattern is more complicated than previously assumed for the Fibonacci sequence. For even REDs the interval is 60, but for odd REDs the intervals can be 20 or 40 which indeed sum to 60, and for Class $\bar{0}_5$ the intervals are 30 (Table 1.3.7).

Finally, the structure of the F_n^2 sequence is

$$\bar{1}_5\,\bar{1}_5\,\bar{4}_5\,\bar{4}_5\,\bar{0}_5\,\bar{4}_5\,\bar{4}_5\,\bar{1}_5\,\bar{1}_5\,\bar{0}_5\,\bar{1}_5\,\bar{1}_5\,\bar{4}_5\,\bar{4}_5\,\bar{0}_5\,\bar{4}_5\,\bar{4}_5$$

which follows from the restricted distribution of the squares in Z_5. This simple structure facilitates the formation of Pythagorean triples from F_n. and known results such as

$$F_{2n-1} = F_n^2 + F_{n-1}^2, n > 1, \quad \text{and} \quad F_{n+2} = 1 + \sum_{j=1}^{n} F_j \,.$$

These can also be related to the structure which we shall do in this book as well as links with primes and the golden section. Mario Livio's book provides a vivid picture of Fibonacci's world, the ramifications of the Fibonacci sequence, and the wider ramifications of the liberating arts! "A curriculum based on great books, ordered by the chronological development of Western civilisation and a treatment of its great themes, approached from the perspectives of multiple disciplines, will ensure that the sense of being part of a shared quest will be genuinely experienced by students and staff, making good conversations and deep inquiry possible. Such programs in Western civilisation within our universities, based on a genuine communal desire to discover ‘the nature of things’ (to recall the title of Lucretius’s great poem), may even be an essential step for any genuine restoration of civic and political discourse in our time” (McInerney 2017).

Chapter 2

FIBONACCI NUMBERS AND STRUCTURE

Integer structure illustrates how primes represented by $4R_1+1$ are equal to a sum of squares. Such primes are in Class $\bar{1}_4 \subset Z_4$, a modular ring. The rows of squares in Z_4 are well defined and this permits equalities for the primes to be derived from the integer structure. These equalities have the forms

$$R_1 = r_1 + r_0$$

where

$$r_1 = 3n(3n \pm 1)\ or\ 2 + 9n'n' + 1)$$

and

$$r_0 = 2^{2q}\left(12m(3m \pm 1) + 1\right)\ or\ 2^{2q}\left(4\left(2 + 9m'\left(m' + 1\right)\right) + 1\right),$$

with q = 0, 1, 2, 3,… and *n, m* yielding the pentagonal numbers, and *n', m'* the triangular numbers. When *n* = *m* the equations are similar to Euler's prime equation. Equations for the remaining primes, in Class $\bar{3}_4$ may be obtained in the same manner using $p = y^2 - x^2$ with $(y - x) = 1$.

An integer structure (IS) analysis of the sum $\left(x^4 + y^4\right)$ is also carried out using the modular ring Z_6. This sum generates many primes and the row structure of such primes is explored. The class functions of the composite factors of this sum are also given, and these, together with the associated row functions, illustrate why it is impossible to produce an integer to the fourth power from such sums. The overall results are consistent with those previously found with IS analysis. Just as the Fibonacci and Mersenne primes are characterised by a two-prime partnership, so too some primes in the modular ring Z_4 depend on the 'row' being prime, particularly in the two odd classes $\bar{1}_4$ and $\bar{3}_4$. Yields of primes from the latter are at least twice those from the former. In effect this seems to occur because $\bar{3}_4$ contains no even powers and so there is a larger interval for primes to appear in any given range.

2.1. Infinite Series and Modular Rings

Thirteen convergent infinite series are here expressed in terms of modular rings. This enables one to assess the contribution of different categories of integers to the infinite series. One class of even integers contributes $\frac{1}{6}\left(\frac{\pi}{4}\right)^2$ to a zeta-function with exponent 2. Another class of even integers makes one quarter the contribution of all the odd integers to this series.

The concept of infinity has intrigued philosophers and mathematicians for thousands of years with questions such as "how can we add an infinity of quantities and arrive at a finite answer?" In fact, many such convergent infinite series have been developed. Probably the one of most current interest to professionals and amateurs alike is the zeta-function:

$$\varsigma(n) = \sum_{x=1}^{\infty} \frac{1}{x^n},\ n > 1. \tag{2.1.1}$$

In particular, the complex zeros of the Riemann zeta-function are presumed to induce the local variations in the distributions of the primes. *Q*-generalizations of the zeta function have been explored in the context of enrichment work by Kim et al. (in press). Hence, this zeta-function has received an enormous amount of attention in order to prove the Riemann Hypothesis which is that all the nontrivial zeros (the values of *n* other than -2,-4,-6,…) of the zeta-function have real part ½; that is, the values of *s* other than -2,-4,-6,… such that $\varsigma(s)$ =0 all lie on the critical line $\sigma = R(s) = \frac{1}{2}$.

"Riemann's 'hypothesis' is the most tantalizing of the unsolved problems of mathematics" (Conway and Guy 1996). It is the number one problem for the 21st century according to Smale (1998). Riemann (1859) developed a clever method for connecting the distribution of primes to properties of the function $\zeta(s)$. Apostol (1980) has an introduction to the relevant analytic number theory, and Edwards (1974) has an exposition of some of the early large-scale calculation attacks on the problem.

In this section we use modular rings to describe some of these series in terms of integer structure so that the composition of the series can be analysed. This opens up exercises and projects for both secondary and tertiary students of mathematics, especially those preparing to become mathematics teachers.

The approach we take is somewhat analogous to that of Effinger (et al 2005) who "contrast and compare the ring of integers and the ring of polynomials in a single variable over a finite field." The notation we adopt is based on the classic text of Hillman and Alexanderson (1973).

Two rings will be used here, namely Z_4 and Z_6, which contain four and six classes respectively.

Table 2.1.1. Square functions for the odd integers

	N^2	Class of N	Function for N	Parity of n
Z_4	$12n(3n+1)+1$ $n=0,1,2,3,...$ $\bar{1}_4$	$\bar{1}_4$ $\bar{3}_4$	$1+12t$ $7+12t$ $t=0,1,2,3,...$	*even* *odd*
	$12n(3n-1)+1$ $n=1,2,3,...$ $\bar{1}_4$	$\bar{1}_4$ $\bar{3}_4$	$5+12t$ $11+12t$	*odd* *even*
	$4(1+3n)(2+3n)+1$ $n=0,1,2,3,...$ $\bar{1}_4$	$\bar{1}_4$ $\bar{3}_4$	$3 \mid N$ $3 \mid N$	*odd* *even*
Z_6	$12n(3n+1)+1$ $\bar{4}_6$	$\bar{4}_6$	$6r_4+1$ $r_4=0,1,2,3,...$	*odd* and *even*
	$12n(3n-1)+1$ $\bar{4}_6$	$\bar{2}_6$	$6r_2-1$ $r_2=0,1,2,3,...$	*odd* and *even*
	$4(1+3n)(2+3n)+1$ $\bar{6}_6$	$\bar{6}_6$	$6r_6+3$ $r_6=0,1,2,3,...$ $3 \mid N$	*odd* and *even*

$\underline{Z_4}$ The integers in this modular ring may be represented by $4r_i+\bar{i}$, in which $\bar{i}$ is the class and r_i can be considered as the row in an array with the four classes as columns. Even integers occur in Classes $\bar{0}_4\,(4r_0)$ and $\bar{2}_4\,(4r_2+2)$ with $r_0 r_2 = 0,1,2,3,...$. There are no powers in Class $\bar{2}_4$. Odd integers occur in Classes $\bar{1}_4\,(4r_1+1)$ and $\bar{3}_4\,(4r_3+3)$ with

$r_1, r_3 = 0,1,2,3,...$. There are no even powers in Class $\bar{3}_4$. Characteristics of the squares of the odd integers are given in Table 2.1.1 (Leyendekkers et al. 1995, 1997). These features make this one of the most useful features and we have used it extensively. In particular we have used it to develop the Erdös-Strauss Conjecture (Shannon and Leyendekkers 2016) and the Brocard-Ramanujan Conjecture (Leyendekkers at all 2017).

Z_6 Integers in this ring are given by $(6r_i + (i-3))$ (Table 2.1.2).

The even integers occur in Classes $\bar{1}_6, \bar{3}_6, \bar{5}_6$. All integers in $\bar{3}_6$ have $3|N$, and there are no even powers in Class $\bar{5}_6$. The odd integers occur in Classes $\bar{2}_6, \bar{4}_6, \bar{6}_6$. All integers in $\bar{6}_6$ have $3|N$, while there are no even powers in Class $\bar{2}_6$.

The classification of the various integers in these rings allows a more detailed analysis of infinite series since the contributions of the different classes can be readily assessed. Some examples which cover various infinite series are now given. Table 2.1.3 lists some convergent infinite series. The aim here is to interpret these in terms of integer structure with the modular rings Z_4 and Z_6.

Table 2.1.2. Integers in Z_6

Class → Row ↓	$\bar{1}_6$	$\bar{2}_6$	$\bar{3}_6$	$\bar{4}_6$	$\bar{5}_6$	$\bar{6}_6$
0	−2	−1	0	1	2	3
1	4	5	6	7	8	9
2	10	11	12	13	14	15
3	16	17	18	19	20	21
4	22	23	24	25	26	27
5	28	29	30	31	32	33

Table 2.1.3. Some convergent infinite series

Type	Series	Value
A	$1/1^2 + 1/2^2 + 1/3^2 + 1/4^2 + \ldots$	$\pi^2/6 = \zeta(2)$
B	$1 - 1/2 + 1/4 - 1/5 + 1/7 - 1/8 + 1/10 - \ldots$	$\pi/3\sqrt{3}$
C	$(1/1^2 \times 1/2^2) + (1/3^2 \times 1/4^2) + (1/5^2 \times 1/6^2) + (1/7^2 \times 1/8^2) + \ldots$	$S(Z_4), S(Z_6)$
D	$\dfrac{8 \times 80 \times 224 \times 440 \times \ldots}{9 \times 81 \times 225 \times 441 \times \ldots}$	$\sqrt{3}/2$
E	$1/2 + 1/4 + 1/8 + 1/16 + 1/32 + 1/64 + \ldots$	1
F	$\dfrac{2 \times 2 \times 4 \times 4 \times 6 \times 6 \times 8 \times 8 \times \ldots}{3 \times 3 \times 5 \times 5 \times 7 \times 7 \times 9 \times 9 \times \ldots}$	$\pi/2$
G	$1 - 1/3 + 1/5 - 1/7 + 1/9 - 1/11 + 1/13 - \ldots$	$\pi/4$
H	$1/(1 \times 2^1) - 1/(3 \times 2^3) + 1/(5 \times 2^5) - 1/(7 \times 2^7) + \ldots$ $+ 1/(1 \times 3^1) - 1/(3 \times 3^3) + 1/(5 \times 3^5) - 1/(7 \times 3^7) + \ldots + \ldots$	$\pi/4$
I	$4\{1/(1 \times 5^1) - 1/(3 \times 5^3) + 1/(5 \times 5^5) - 1/(7 \times 5^7) + \ldots\}$ $- \{1/(1 \times 239^1) - 1/(3 \times 239^3) + 1/(5 \times 239^5) - 1/(7 \times 239^7) + \ldots\} + \ldots$	$\pi/4$

Type	Series	Value
J	$1/(1\times 3)+1/(5\times 7)+1/(9\times 11)+1/(13\times 15)+\ldots$	$\pi/8$
K	$1/2+1/12+1/30+1/56+1/90+\ldots$	$\ln 2$
L	$3\times\frac{36}{35}\times\frac{144}{143}\times\frac{324}{323}\times\frac{576}{575}\times\ldots$	π
M	$1/(2\times 3)+1/(3\times 4)+1/(4\times 5)+1/(5\times 6)+1/(6\times 7)+\ldots$	½

Series A: Using Z_4 we can distinguish among the sums for the different classes of integers.

In $\bar{0}_4$, the integers are represented by $N = 4r_0$, so that $N^2 = 16r_0^2$, and the sum becomes

$$S(\bar{0}_4) = \frac{1}{16}\sum_{r_0=1}^{\infty}\frac{1}{r_0^2}$$
$$= \frac{1}{16}\zeta(2)$$
$$= \frac{1}{6}\left(\frac{\pi}{4}\right)^2.$$

The result shows that integers in this class are linked to the Euler and Machin Series (H and I respectively in Table 2.1.3).

In $\bar{2}_4$, $N = 4r_2 + 2$, and $N^2 = 4(2r_2 + 1)^2$, so that the sum becomes

$$S(\bar{2}_4) = \frac{1}{4}\sum_{r=0}^{\infty}\frac{1}{(2r+1)^2}$$

which is one quarter the sum of the corresponding odd integers in Classes $\bar{1}_4$ and $\bar{3}_4$.

In $\bar{1}_4$, $N = 4r_1 + 1$ and $N^2 = 4R_1 + 1$ and the squares have three functions (Table 2.1.1). For $\bar{1}_4$ these are

$$12n(3n+1)+1, \ (n = 0,2,4,6,\ldots),$$
$$12n(3n-1)+1, \ (n = 1,3,5,7,\ldots),$$
$$4(3n+1)(3n+2)+1, \ 3 \mid N, \ (n = 1,3,5,7,\ldots).$$

Hence the sum is

$$S(\bar{1}_4)=\sum_{n=0}^{\infty}\frac{1}{24n(6n+1)+1}+\frac{1}{24(2n+1)(3n+1)+1}+\frac{1}{8(3n+2)(6n+5)+1}.$$

Of course, the simple sum

$$S(\bar{1}_4)=\sum_{r=0}^{\infty}\frac{1}{(4r+1)^2}$$

also applies, but this fails to discriminate among the different types of integers within this class (Table 2.1.1).

In $\bar{3}_4$, $N=4r_3+3$, but $N^2=4R_1+1$, as there are no even powers in this class. As in $\bar{1}_4$, the squares follow the three functions in Table 1 with reversed parity for *n*.

$$S(\bar{3}_4)=\sum_{n=0}^{\infty}\frac{1}{12(2n-1)(6n-2)+1}+\frac{1}{24(n+1)(6n+5)+1}+\frac{1}{8(6n+1)(3n+1)+1}.$$

As for class $\bar{1}_4$, there is a simple but restrictive sum:

$$S(\bar{3}_4)=\sum_{r=0}^{\infty}\frac{1}{(4r+3)^2}.$$

As can be seen from Table 2.1.1, the class structure in Z_6 is simpler than in Z_4, at least for the squares, so that only one function applies for each of the three classes of odd integers.

Series B: This sum obviously excludes the integers *N* such that $3|N$. Hence, the modular ring Z_6 is the most appropriate for any analysis, with integers of type $(6r+3)\in\bar{6}_6$ and $6r_3\in\bar{3}_6$ being excluded and with

$(1/N_i)$ following the pattern $\bar{1}_6\bar{2}_6\bar{4}_6\bar{5}_6$ for N; that is, each set of consecutive four-sum-components may be represented by

$$\frac{1}{(6r_1-2)}-\frac{1}{6r_2-1}+\frac{1}{6r_4+1}-\frac{1}{6r_5+2}$$

and, since $r_1=r_2=r_4=r_5$, the series may be expressed as

$$\frac{\pi}{3\sqrt{3}}=\frac{1}{2}+\sum_{r=1}^{\infty}\frac{18r^2+1}{(9r^2-1)(36r^2-1)}.$$

Series C: For Z_4 we have the Class sequence

$$\left\{\left(\frac{1}{\bar{1}_4^2\times\bar{2}_4^2}\right)+\left(\frac{1}{\bar{3}_4^2\times\bar{0}_4^2}\right)\right\}$$ repeated with $r=r_2=r_3=r_4$

and $r_0=r+1$, so that

$$S(Z_4)=\sum_{r=0}^{\infty}\frac{1}{(4r+1)^2\times(4r+2)^2}+\frac{1}{(4r+3)^2\times(4r+4)^2}.$$

For Z_6 we have the Class sequence

$$\left\{\left(\frac{1}{\bar{4}_6^2\times\bar{5}_6^2}\right)+\left(\frac{1}{\bar{6}_6^2\times\bar{1}_6^2}\right)+\left(\frac{1}{\bar{2}_6^2\times\bar{3}_6^2}\right)\right\}$$ repeated with $r=r_4=r_5=r_6$

and $r_1=r_2=r_3=r+1$, so that

$$S(Z_6) = \sum_{r=0}^{\infty} \frac{1}{(6r+1)^2 \times (6r+2)^2} + \frac{1}{(6r+3)^2 \times (6r+4)^2} + \frac{1}{(6r+5)^2 \times (6r+6)^2}.$$

Series D: For Z_4, this becomes

$$\frac{\sqrt{3}}{2} = \prod_{n=0}^{\infty} \frac{4(3n+1)(3n+2)}{4(3n+1)(3n+2)+1}$$

since $3|N$ and $2 \nmid N$.

For Z_6, this becomes

$$\frac{\sqrt{3}}{2} = \prod_{r=0}^{\infty} \frac{(6r+3)^2 - 1}{(6r+3)^2}$$

since $\bar{6}_6$ only is involved.

Series E: Here $\frac{1}{N}, N \in \bar{0}_4 \subset Z_4$, has each successive row double the previous one, and so the sum is

$$S'(\bar{0}_4) = \frac{1}{2} + \sum_{j=0}^{\infty} \frac{1}{2^{j+2}}.$$

Series F: In terms of the Z_4 class structure

$$\frac{\pi}{2} = \frac{\bar{2}_4 \bar{2}_4 \bar{0}_4 \bar{0}_4 \bar{2}_4 \bar{2}_4 \bar{0}_4 \bar{0}_4}{\bar{1}_4 \bar{3}_4 \bar{3}_4 \bar{1}_4 \bar{1}_4 \bar{3}_4 \bar{3}_4 \bar{1}_4}, \text{ so that}$$

$$\frac{\pi}{2} = \prod_{r=0}^{\aleph} \frac{(4r+2)^2 (4r+4)^2}{(4r+1)^2 (4r+3)^2 (4r+5)^2}.$$

For Z_6, the structure is $\dfrac{\overline{5}_6\overline{5}_6\overline{1}_6\overline{1}_6\overline{3}_6\overline{3}_6}{\overline{4}_6\overline{6}_6\overline{6}_6\overline{2}_6\overline{2}_6\overline{4}_6}$

with $r_4 = r_5 = r_6 = r;\ r_1 = r_2 = r_3 = r_4 = r+1.$

Series G: For Z_4 this sum becomes

$$S(Z_4) = \frac{\pi}{4}$$

$$= 2\sum_{r=0}^{\infty}\frac{1}{(4r+1)(4r+3)}\quad .$$

$$= 1 - 2\sum_{r=0}^{\infty}\frac{1}{(4r+3)(4r+5)}.$$

Series H and I: These may be expressed as $\left(6S\left(\overline{0}_4\right)\right)^{1/2}$. Readers might like to express these series in terms of the rows $r \in \{Z_4, Z_6\}$.

***Series J*:** This series is actually Series G. If we use Z_6 we get

$$\frac{\pi}{8} = \frac{1}{3} + \sum_{r=1}^{\infty}\frac{1}{\left(36(2r-1)^2-1\right)} + \frac{1}{(12r-1)(12r-3)} + \frac{1}{(12r+3)(12r+1)}.$$

Series K: This series is related to Series C. Both are specific cases of the more general

$$S = \frac{1}{1^n \times 2^n} + \frac{1}{3^n \times 4^n} + \frac{1}{5^n \times 6^n} + \dots\ .$$

The class structure in Z_4 here has the form

$$\frac{1}{\bar{1}_4 \times \bar{2}_4} + \frac{1}{\bar{3}_4 \times \bar{0}_4}$$

which is repeated with changing row, and so

$$\ln 2 = \sum_{r=0}^{\infty} \frac{1}{(4r+1)(4r+2)} + \frac{1}{(4r+3)(4r+4)}.$$

Series L: Here all the elements of the numerator belong to $\bar{3}_6 \in Z_6$. Hence

$$\pi = 3\prod_{r=0}^{\infty} \frac{36r^2}{36r^2 - 1}.$$

This can be compared with Series F and its more complicated structure.

Series M: The class structure here is

$$\frac{1}{\bar{2}_4 \bar{3}_4} + \frac{1}{\bar{3}_4 \bar{0}_4} + \frac{1}{\bar{0}_4 \bar{1}_4} + \frac{1}{\bar{1}_4 \bar{2}_4}$$

with $r_0 = r_1 = r_2' = r,\ r_2 = r_3 = r - 1;$

$$\frac{1}{2} = \sum_{r=1}^{\infty} \frac{1}{4r-1}\left(\frac{1}{4(r-1)+2} + \frac{1}{4r}\right) + \frac{1}{4r+1}\left(\frac{1}{4r+2} + \frac{1}{4r}\right)$$
$$= \sum_{r=1}^{\infty} \frac{1}{4r^2 - 1}$$

which can be compared with E.

2.2. Equations for Primes Obtained from Integer Structure[2]

Fermat established that only primes of the form $4r_1+1$ (in our notation) equalling a sum of squares. This follows very simply from the integer structure. We use the modular ring Z_4 since $4r_1+1$ represents the function for primes in class $\overline{1}_4$ of Z_4 (Table 2.2.1).

All powers of even integers fall in Class $\overline{0}_4$ whilst all even powers of odd integers fall in Class $\overline{1}_4$ [1]. Hence, with x odd and y even,

$$x^2+y^2=\overline{1}_4+\overline{0}_4=\overline{1}_4 \tag{2.2.1}$$

so that all odd integers equalling a sum of squares must fall in Class $\overline{1}_4$. However, for composites in Class $\overline{1}_4$, all do not equal a sum of squares, as shall be shown below.

Table 2.2.1. First seven elements of classes modulo 4

Function		$4r_0$	$4r_1+1$	$4r_2+2$	$4r_3+3$
Class		$\overline{0}_4$	$\overline{1}_4$	$\overline{2}_4$	$\overline{3}_4$
Row	0	0	1	2	3
	1	4	5	6	7
	2	8	9	10	11
	3	12	13	14	15
	4	16	17	18	19
	5	20	21	22	23
	6	24	25	26	27

[2] Based on Leyendekkers and Shannon (2010).

(i) Right-end-digit (RED) Analysis

To simplify the calculations we use REDs. An asterisk indicates the RED, N represents the integer (Table 2.2.2).

Table 2.2.2. RED analyses

N^*	x^*	y^*
1	1,5,9	0,4,6
3	3,7	2,8
7	1,9	4,6
9	3,5,7	0,2,8

As an example, consider $N^* = 7$, then $x = 1, 11, 21, 31, 41, \ldots$ or $x = 9, 19, 29, 39, 49, \ldots$

Let

$$N - x^2 = y^2 \qquad (2.2.2)$$

or in RED terms, 7* – 1* = 6* , then if $N - x^2$ is a square a value for y has been found and $(y^2)^* = 6^*$.

(ii) Composites versus Primes

Examples for composites in Class $\overline{1}_4$ and with $N^* = 7$, are listed in Table 2.2.3.

Only relatively few composites equal a sum of squares. As can be seen, those that do have factors in Class $\overline{1}_4$. Factors in Class $\overline{3}_4$ inhibit N from 31equalling a sum of squares. Note that the factors in $\overline{3}_4$ must be even since N must be in $\overline{1}_4$. That is, $\overline{3}_4\,\overline{3}_4 \subset \overline{1}_4$ or $\overline{3}_4\,\overline{3}_4\,\overline{3}_4\,\overline{3}_4 \subset \overline{1}_4$ but $\overline{3}_4\,\overline{3}_4\,\overline{3}_4 \subset 3_4$, and so on.

Table 2.2.3. Classes of factors

N	Row	(x,y)	Factors	Class of factors
57	14	-	3 ×19	$\bar{3}_4\,\bar{3}_4$
77	19	-	11×7	$\bar{3}_4\,\bar{3}_4$
117	29	(99,6)	9×13	$1_4\ 1_4$
177	44	-	3×59	$\bar{3}_4\,\bar{3}_4$
217	54	-	7×31	$\bar{3}_4\,\bar{3}_4$
237	59	-	3×79	$\bar{3}_4\,\bar{3}_4$
297	74	-	3×11	$\bar{3}_4\,\bar{3}_4$
357	89	-	3×119	$\bar{3}_4\,\bar{3}_4$
377	94	(11,16)(19,4)	13×29	$1_4\ 1_4$
417	104	-	3×139	$\bar{3}_4\,\bar{3}_4$
437	109	-	23×19	$\bar{3}_4\,\bar{3}_4$
477	119	(21,6)	3×53	$1_4\ 1_4$
497	124	-	7×71	$\bar{3}_4\,\bar{3}_4$
517	129	-	11×47	$\bar{3}_4\,\bar{3}_4$
537	134	-	3×179	$\bar{3}_4\,\bar{3}_4$
597	149	-	3×199	$\bar{3}_4\,\bar{3}_4$
637	159	-	3×53	$\bar{3}_4\,\bar{3}_4$
657	164	(9,24)	9×73	$1_4\ 1_4$
697	174	(11,24)(21,16)	17×41	$1_4\ 1_4$
717	179	-	3×239	$\bar{3}_4\,\bar{3}_4$
737	184	-	11×67	$\bar{3}_4\,\bar{3}_4$

N	Row	(x,y)	Factors	Class of factors
777	194	-	3×7×37	$\overline{3}_4\,\overline{3}_4\,1_4$
817	204	-	19×43	$\overline{3}_4\,\overline{3}_4$
837	209	-	3×31	$\overline{3}_4\,\overline{3}_4$
897	224	-	3×13×23	$\overline{3}_4\,1_4\,\overline{3}_4$
917	229	-	7×131	$\overline{3}_4\,\overline{3}_4$
957	239	-	3×79	$\overline{3}_4\,\overline{3}_4$
1017	254	(21,24)	9×113	$1_4\;1_4$
1037	259	(29,14)(19,26)	17×61	$1_4\;1_4$
1057	264	-	7×151	$\overline{3}_4\,\overline{3}_4$
1157	289	(31,14)(1,34)	13×89	$1_4\;1_4$

When only a unique (*x*, *y*) occurs *x* and *y* have a common factor. Otherwise multiple values of (*x*, *y*) occur. These characteristics distinguish the composites from primes which only have a unique set of (*x*, *y*) (Table 2.2.4). This means that an integer in $\overline{1}_4$ may be checked for primality using the $(x^2 + y^2)$ characteristics.

Table 2.2.5 shows the possible number of trials for x that are needed. These are often less than the number of prime factors needed to establish primality. As well, additional information, that is, (*x*, *y*) is obtained. Another advantage is that no primes are needed for testing the primality of *N*. However, more importantly, the aim here is to establish equations that derive from integer structure and hence more accurately describe primes.

(iii) Characteristics of different N^*

Only $N^* = 1$ or 7 gives $x = 1$. For this value of x the row of *N* must be a square since

$$4r_1 + 1 = 1 + y^2 \tag{2.2.2}$$

Thus $r_1 = (y/2)^2$

Table 2.2.4. Values of variables

p	Row	(x,y)	p	Row	(x,y)	p	Row	(x,y)	p	row	(x,y)
5	1	1,2	269	67	13,10	593	148	23,8	941	235	29,10
13	3	3,2	277	69	9,14	601	150	5,24	953	238	13,28
17	4	1,4	281	70	5,16	613	153	17,18	977	244	31,4
29	7	5,2	293	73	17,2	617	154	19,16	997	249	31,6
37	9	1,6	313	78	13,12	641	160	25,4	1009	252	15.28
41	10	5,4	317	79	11,14	661	165	25,6	1013	253	23,22
53	13	7,2	337	84	9,16	673	168	23,12	1021	255	11,30
61	15	5,6	349	87	5,18	677	169	1,26	1033	258	3,32
73	18	3,8	353	88	17,8	701	175	5,26	1049	262	5,32
89	22	5,8	373	93	7,18	709	177	15,22	1061	265	31,10
97	24	9,4	389	97	17,10	733	183	27,2	1069	267	13,30
101	25	1,10	397	99	19,6	757	189	9,26	1093	273	33,2
109	27	3,10	401	100	1,20	761	190	19,20	1097	274	29,16
113	28	7,8	409	102	3,20	769	192	25,12	1109	277	25,22
137	34	11,4	421	105	15,14	773	193	17,22	1117	279	29,26
149	37	7,10	433	108	17,12	797	199	11,26	1129	282	27,20
157	39	11,6	449	112	7,20	809	202	5,28	1153	288	33,8
173	43	13,2	457	114	21,4	821	205	25,14	1181	295	5,34
181	45	9,10	461	115	19,10	829	207	27,10	1193	298	13,32
193	48	7,12	509	127	5,22	853	213	23,18	1201	300	25,24
197	49	1,14	521	130	11,20	857	214	29,4	1213	303	27,22
229	57	15,2	541	135	21,10	877	219	29,6	1217	304	31,16
233	58	13,8	557	139	19,14	881	220	25,16	1229	307	35,2
241	60	15,4	569	142	13,20	929	232	23,20	1237	309	9,34
57	64	1,16	577	144	1,24	937	234	19,24	1249	312	15,32

Table 2.2.5. Trials of variable

p	row	x, y	Number of trials of x
1277	319	11,34	1,11 (9,19,29)
1289	322	35,8	5,25,35 (3,13,23,33) (7,17,27)
1297	324	1,36	1, (9,19,29)
1301	325	25,26	5,25 (1,11,21,31) (9,19,29)
1321	330	5,36	5 (1,11,21,31) (9,19,29)
1361	340	31,20	1,11,21,31 (5,15,25,35) (9,19,29)
1373	343	37,2	7,17,27,37 (3,13,23,33)
1409	352	25,28	3,13,23,33 (5,15,25) (7,17,27,37)
1597	399	21,34	1,11,21 (9,19,29,39)
1613	403	13,38	3,13 (7,17,27,37)
1621	405	39,10	1,11,21,31 (5,15,25,35) (9,19,29,39)
1637	409	31,26	1,11,21,31 (9,19,29,39)
1657	414	19,36	9,19 (1,11,21,31)

For $p^* = 7$, when $(y/2)^2$ is even the row of the row is also a square. When $(y/2)^2$ is odd then the row of the row = 6*K*, with

$$K = \frac{n}{2}(3n \pm 1) \qquad (2.2.3)$$

provided $3 \nmid \frac{y}{2}$ [1]. If $3 | \frac{y}{2}$ the row of $(y/2)^2$ follows the triangular numbers.

For $p^* = 1$ and $x = 1$, $5 | r_1$. When $(y/2)^2$ is even, the row of the row is square. When $(y/2)^2$ is odd the row of the row = 6*K*, as for $p^* = 7$.

For $p^* = 9$ the rows of the primes have REDs 2 or 7 hence $r_i - r_{i-1} = 5k$. Since $x^* = 5$ dominates it would be best to start with the RED, i.e. 5, 15, 25, 35,… when searching for the appropriate x.

For $p^*=3$ the rows of the primes have REDs of 3 or 8, thus $r_i - r_{i-1} = 5k$. There is an even distribution between $x^*=3$ and $x^*=7$.

As an example, we shall examine the row structure for $p^*=7$. This will indicate the sort of functions that are most appropriate for predicting primes. For these primes $x^*=1,9$ and $y^*=4,6$ (Table 2.2.2). With

$$x^2 = 4r_1 + 1 \tag{2.2.4}$$

Then $r_1 = 6K$ with

$$K = \tfrac{n}{2}(3n \pm 1) \tag{2.2.5}$$

provided $3 \nmid x$ [1]. When $3 \mid x$

$$r_1 = 2 + 18\sum_{i=1}^{j} i \tag{2.2.6}$$

with

$$\sum_{i=1}^{j} i = \tfrac{n}{2}(n+1) \tag{2.2.7}$$

Also

$$y^2 = 4r_0 \tag{2.2.8}$$

with r_0 being a square, that is

$$r_0 = (y/2)^2 \tag{2.2.9}$$

The row of r_0 (or r_1') when $(y/2)$ is odd, is given by $r_1' = 6K$ as above or, when having a factor of 3, r_1' is given by equation (2.2.4). When $(y/2)$ is even the row will be a square.

(iv) $(y/2)$ is odd

***(a) x, y* prime to 3. With**

$$x^2 = 4r_1 + 1 \tag{2.2.10}$$

Then $r_1 = 3n(3n \pm 1)$ and $y^2 = 4r_0$, r_0 odd (2.2.11)

$$r_0 = (y/2)^2 = 4r_1' + 1 \tag{2.2.12}$$

with $r' = 3m(3m \pm 1)$.

Then

$$\begin{aligned} p &= x^2 + y^2 \\ &= 4r_1 + 1 + 4r_0 \\ &= 4(r_1 + r_0) + 1 \end{aligned} \tag{2.2.13}$$

The row of the prime, R_1, equals $(r_1 + r_0)$ thus

$$R_1 = 3n(3n \pm 1) + 12m(3m \pm 1) + 1 \tag{2.2.14}$$

Let $\frac{q}{2}(3q+1) \in \mathrm{I}$, $q = n, m$ and $\frac{q}{2}(3q-1) \in \mathrm{II}$, then there will be four different combinations to give the equations for rows of primes and hence the prime from $4r_1 + 1$ (Table 2.2.6).

It is found that *x*, *y*, *n* and *m* follow regular functions (Table 2.2.7). These are indicated in Table 2.2.8 illustrating prime production from the $f(n,m)$ in Table 2.2.6.

Table 2.2.6. Functions yielding prime rows

f(*n*) category for *x*, *y*	Functions giving rows of primes	No
I.I	$9(n^2+4m^2)+3(n+4m)+1$	1
II.II	$9(n^2+4m^2)-3(n+4m)+1$	2
II.I	$9(n^2+4m^2)-3(n-4m)+1$	3
I.II	$9(n^2+4m^2)+3(n-4m)+1$	4

Table 2.2.7. Functions for x, y and $f(q)$ for rows of x^2 and $(y/2)^2$ (odd)

$f(q)$	x	n	$(y/2)$ odd	m
I	19 + 30t	3 + 5t (a)	7 + 30s	1 + 5s I
	31 + 30t	5 + 5t (b)	13 + 30s	2 + 5s (f)
II	11 + 30t	2 + 5t I	17 + 30s	3 + 5s (g)
	29 + 30t	5 + 5t (d)	23 + 30s	4 + 5s (h)

Table 2.2.8. Equations yielding regular functions

y	x	n	t	f(t)	m	a	f(s)	R_1 row of p	Prime, p, $4R_1+1=x^2+y^2$
I.I Equation 1 Table 2.2.6									
14	19	3	0	a	1	0	e	139	557
	61	10	1	b				979	3917
206	19	3	0	a	17	3	f	10699	42797
	31	5	0	b				10849	43397
	79	13	2	a				12169	48677

y	x	n	t	f(t)	m	a	f(s)	R_1 row of p	Prime, p, $4R_1+1=x^2+y^2$
314	61	10	1	b	26	5	e	25579	102317
	91	15	2	b				26719	106877
	109	18	3	a				27619	110477
II.II Equation 2 Table 2.2.6									
34	11	2	0	c	3	0	g	319	1277
	41	7	1	c				709	2837
	59	10	1	d				1159	4637
	71	12	2	c				1549	6197
214	59	10	1	d	18	3	g	12319	49277
	89	15	2	d				13429	53717
	101	17	3	c				13999	55997
346	29	5	0	d	29	5	h	30139	120557
	89	15	2	d				31909	127637
	101	17	3	c				32479	129917
II.I Equation 3 Table 2.2.6									
26	11	2	0	c	2	0	f	199	797
	41	7	1	c				589	2357
	59	10	1	d				1039	4157
	71	12	2	c				1429	5717
	89	15	2	d				2149	8597
326	11	2	0	c	27	5	f	26599	106397
	71	12	2	c				27829	111317
	89	15	2	d				28549	114197
I.II Equation 4 Table 2.2.6									
34	31	5	0	b	3	0	g	529	2117
	49	8	1	a				889	3557
	61	10	1	b				1219	4877
	91	15	2	b				2359	9437
	109	18	3	a				3259	13037
346	19	3	0	a	29	5	h	30019	120077
	31	5	0	b				30169	120667
	49	8	1	a				30529	122117
	91	15	2	b				31999	127997

(b) x or y with a factor of 3

When $3 \nmid x, r_1 = 3n(3n \pm 1)$ as before, but when $3|y$, r' for r_0 or $(y/2)$ is given by Equation (3.3). Some examples are given in Table 2.2.9. Thus the equations for the rows of the primes are:

$$R_1 = 9(n^2 + 4m^2) + 3(n + 12m) + 9, \quad Ia, \tag{2.2.15}$$

or

$$R_1 = 9(n^2 + 4m^2) - 3(n - 12m) + 9, \quad IIa. \tag{2.2.16}$$

Table 2.2.9. Factored by 3

$(y/2)$	Row of $(y/2)^2$	m	93	2162	15
3	2	0	117	3422	19
27	182	4	123	3782	20
33	272	5	147	5402	24
57	812	9	153	5852	25
63	992	10	177	7832	29
87	1892	14→↑	183	8372	30

Examples using these equations are listed in Table 2.2.10.

When $3|x$ we get

$$\begin{aligned} r_1 &= 2 + 18\left(\tfrac{1}{2}n(n+1)\right) \\ r_0 &= 4r_1' + 1 \end{aligned} \tag{2.2.17}$$

with

$$r' = 3m(3m \pm 1), \tag{2.2.18}$$

so row of prime, R_1, is given by

$$R_1 = 9\left(n^2 + 4m^2\right) + 3(3n + 4m) + 3,\ m \in I, \quad (2.2.19)$$

or

$$R_1 = 9\left(n^2 + 4m^2\right) + 3(3n - 4m) + 3,\ m \in II. \quad (2.2.20)$$

For example, let $x = 9$, $y = 14$, then $r_1 = 20$, $n = 1$ and row = 49, $r' = 6\times2$, $m = 1$ (I) so that row of prime = 69 and prime = 277.

Table 2.2.10. Examples of equations (2.2.15; 2.2.16)

x	y	Row of x^2	Row of $(y/2)^2$	n	Eq	m	Row of p	Prime, p
11	6	30	2	2	II	0	39	157
19		90	2	3	I	0	99	397
29		210	2	5	II	0	219	877
31		240	2	5	I	0	249	997
91		2070	2	15	I	0	2079	8317
11	234	30	3422	2	II	19	13719	54877
19		90	3422	3	I	19	13779	55117
31		240	3422	5	I	19	13929	55717
41		420	3422	7	II	19	14109	56437
59		870	3422	10	II	19	14559	58237
61		930	3422	10	I	19	14619	58477
71		1260	3422	12	I	19	14949	59797

(c) **$(y/2)$ is even**

In general, $(y/2)^2 = r_0$ and $(y/2) = 2^q t$ where t is a positive odd integer.

When $q = 0$, then $y/2$ will be odd, so that $r_0 = t^2$ (Section 3(i).)

Here $q > 0$ and

$$r_0 = 2^{2q} t^2. \quad (2.2.21)$$

***(d)* $t > 1$ and $3 \nmid t, x$.**

The same form of equation as above will also apply here, that is with prime $p \in \bar{1}_4$

$$p = 4r_1 + 1 = x^2 + y^2 \tag{2.2.22}$$

with $x^2 = 4r_1' + 1$ and $y^2 = 4r_0$ so that $p = 4(r_1' + r_0) + 1$. (2.2.23)

Then the row of the prime equals $(r_1' + r_0)$ as before.

Since

$$r' = 6K = 3n(3n \pm 1) \tag{2.2.24}$$

and r_0 is represented by

$$\begin{aligned} r_0 &= 2^{2q} t^2 \\ &= 2^{2q}(4r_1'' + 1) \end{aligned} \tag{2.2.25}$$

with $r_1'' = 3m(3m \pm 1)$, the row of the prime becomes

$$r_1' + r_0 = 3n(3n \pm 1) + 2^{2q}(12m(3m \pm 1) + 1). \tag{2.2.26}$$

Some examples are given in Table 2.2.11.

***(e)* x or y with factor of 3**

The development is similar to that in Section 3 (i) (b), except that

$$r_0 = 2^{2q} f(m). \tag{2.2.27}$$

$t = 1$

Table 2.2.11. Examples of (2.2.26) * underlined value used for *p*

y/2	*q*	*t*	r_1''	*K*	*m*	*n* for various *x**	Row of *p*	*p*
22	1	11	30	5	211	I 3,$\underline{5}$,7,8,10	724	2897
						II 5,10,12,15		
52	2	13	42	7	21	I 3,5,7,8,10	2734	10937
						II $\underline{2}$,5		
62	1	31	240	40	51	I 3,13	6394	25577
						II 2,5,15,$\underline{17}$		
82	1	41	420	70	711	I $\underline{8}$	7324	29297
						II 2,5,17		
92	2	23	132	22	411	I 3,7,$\underline{18}$	11434	45737
						II 10,15		

The row of the prime is simply given by

$$R_1 = r_1' + r_0 = 3n(3n \pm 1) + 2^{2q}. \qquad (2.2.28)$$

For example, with $y/2 = 128$, $r_0 = 16384$, $q = 7$, and with $x = 79$, $x^2 = 6241$ with a row $= 6\times260$ so that $n = 131$. Some other suitable x are 11, 29, 41, 61, 109.

Thus, integer structure analysis shows that primes $\in \bar{1}_4$ have the form $4R_1 + 1$ where $R_1 = r_1 + r_0$ from $x^2 = 4r_1 + 1$ and $y^2 = 4r_0$ with $r_1 = 3n(3n \pm 1)$, or if $3|x$, $r_1 = 2 + 9n'(n' + 1)$, and

$$r_0 = 2^{2q}\left(12m(3m \pm 1) + 1\right),$$

or if $3|y$,

$$r_0 = 2^{2q}\left(4\left(2 + 9m'(m'+1)\right) + 1\right).$$

When $q = 0$, $y/2$ is odd, and $m = 0$ yields $r_0 = 2^{2q}$. Primes with a RED of 7 have been used as an example here but the same equations can be applied to all primes in Class $\bar{1}_4$. A similar analysis may be made for primes in Class $\bar{3}_4$ but in this case

$$p = y^2 - x^2 \tag{2.2.29}$$

with $y - x = 1$ and $(y,x) \in \{(\bar{0}_4,\bar{3}_4),(\bar{2}_4,\bar{1}_4)\}$. Since the squares follow $f(n)$ and $f(m)$, similar equations arise. When $n = m$, the equations have the form

$$f(n) = An^2 + Bn + C \tag{2.2.30}$$

which is similar to Euler's prime generating equation, and the recently developed prime equations for the modular ring Z_6 (Leyendekkers et al 2007, 2008).

2.3. Integer Structure Analysis of Primes and Composites from Sums of Two Fourth Powers[3]

Sums of even powers, m, can produce primes, as was shown by Fermat with $2^{2^n} + 1^{2^n}$ $(n \leq 4)$, whereas sums of odd powers may be factored. For example, the Diophantine factorizations:

$$\begin{aligned} x^3 + y^3 &= (x+y)(x^2 + y^2 - xy) \\ x^5 + y^5 &= (x+y)(x^4 + y^4 - xy(x^2 - xy + y^2)) \\ x^7 + y^7 &= (x+y)(x^6 + y^6 - xy(x^4 - xy(x^2 - xy + y^2) + y^4)). \end{aligned} \qquad (2.3.1), (2.3.2), (2.3.3)$$

However, if $\frac{1}{2}m$ is odd, a factored form can arise; for instance,

$$x^6 + y^6 = (x^2 + y^2)((x^2 + 3xy + y^2)^2 - 6xy(x+y)^2). \tag{2.3.4}$$

---

[3] Based on Leyendekkers and Shannon (2007).

This does not happen for the sum of two squares, those, a matter of related interest, Adler (2006) has proved that

$$x^2 + y^2 = 3xy \mp 1$$

generates the Fibonacci numbers with odd and even indices, respectively, by equating x to any Fibonacci number with odd (even) index and solving for y; the larger root will be the Fibonacci number with the next larger odd (even) index.

More particularly, $(x^4 + y^4)$ produces relatively large numbers of primes and in this section we shall examine the underlying structure of this sum by using the modular ring Z_6 (Leyendekkers et al. 1995).

Integers in this ring have the form $(6r_i + (i-3))$ in which the row $r_i \in \bar{i}_6$, the class. Even integers $\in \{\bar{1}_6, \bar{3}_6, \bar{5}_6\}$; $\bar{5}_6$ has no even powers, and $3|N \forall N \in \bar{3}_6$. Odd integers $\in \{\bar{2}_6, \bar{4}_6, \bar{6}_6\}$; $\bar{2}_6$ has no even powers, and $3|N \forall N \in \bar{6}_6$.

Fermat showed that primes generated by $(4r_1 + 1) \in \bar{1}_4 \subset Z_4$ are equal to a unique sum of squares (Leyendekkers et al. 1997). In Z_6, these primes fall in $\bar{2}_6$ in odd rows and in $\bar{4}_6$ in even rows (Leyendekkers and Shannon 2004). Since

$$x^4 + y^4 = (x^2)^2 + (y^2)^2, \tag{2.3.5}$$

the primes formed from Equation (2.3.5) will be of this type. Table 2.3.1 shows that the (x^4, y^4) couple $(\bar{1}_6 \bar{4}_6)$ yields the most primes. This is understandable since $x \in \{\bar{1}_6, \bar{5}_6\}$ and $y \in \{\bar{4}_6, \bar{2}_6\}$ contribute to $(x^4 + y^4)$ which must belong to $\bar{2}_6$ (Table 2.3.2). Furthermore, since $\bar{2}_6$

has no even powers there is room for more primes, whereas $3 \mid N \in \bar{6}_6$ so there are no primes in this class.

Table 2.3.1. (*x*,*y*) couples which produce primes

(*NB*: $\left(\bar{5}_6\right)^4 \in \bar{1}_6 ; \left(\bar{2}_6\right)^4 \in \bar{4}_6$)

$X\downarrow y\rightarrow$	1	3	5	7	9	11	13	15	17	19	21	23	25	27
2	⊗	⊕	⊗	⊗	⊕	⊗			⊗	⊗		⊗		⊕
4	⊗	⊕	⊗	⊗		⊗	⊗		⊗			⊗		
6	*			*		*				*			*	
8		⊕	⊗		⊕			⊕		⊗		⊗	⊗	
10				⊗	⊕		⊗			⊗	⊕	⊗		
12							*			*			*	
14			⊗				⊗	⊕	⊗	⊗				
16	⊗	⊕	⊗			⊗			⊗		⊕			⊕
18			*	*		*						*		
20	⊗	⊕			⊗				⊗					

Legend: $\left(x^4 y^4\right): \bar{1}_6 \bar{4}_6$ ⊗; $\bar{1}_6 \bar{6}_6$ ⊕; $\bar{3}_6 \bar{4}_6$ *.

Table 2.3.2. Classes for $\left(x^4 + y^4\right)$

x^4	y^4	$x^4 + y^4$
$\bar{1}_6$	$\bar{4}_6$ $\bar{6}_6$	$\bar{2}_6$ $\bar{4}_6$
$\bar{3}_6$	$\bar{4}_6$ $\bar{6}_6$	$\bar{4}_6$ $\bar{6}_6$

We may use these functions here to show how the row structure of the primes arises. As noted above, primes in Z_6 equal a sum of squares when they have odd rows in $\bar{2}_6$ and even rows in $\bar{4}_6$. The (*a*,*b*) couples

$(\bar{1}_6, \bar{4}_6)$, $a = x^2, b = y^2$, need only be considered when $3 \nmid N$ (⊗ in Table 2.3.1). These primes fall in odd rows in $\bar{2}_6$. The component $a = x^2 = 2^m Q$ (Table 2.3.3).

Table 2.3.3 shows that *m* is even and $Q \in \bar{4}_6$ so that

$$x^4 = a^2 = 2^{2m}(6r_4 + 1)^2 \tag{2.3.6}$$

and R_1, the row for x^4, is given by

$$R_1 = 2^{2m}\left(6n_2^2 + 2n_2 + \tfrac{1}{6}\right) + \tfrac{1}{3} \tag{2.3.7}$$

with n_2 equal to the row, r_4, of $Q \in \bar{4}_6$ (Table 2.3.3). For example, with a=64= $2^6 \times 1$, m=6, n_2=0, so $R_1 = 683$.

Table 2.3.3. $a = x^2 = 2^m Q$

x	$a = x^2$	m	Q	Class of Q	r_4, row of Q	R_1, row of a^2
2	4	2	1	$\bar{4}_6$	0	3
4	16	4	1	$\bar{4}_6$	0	43
8	64	6	1	$\bar{4}_6$	0	683
10	100	2	25	$\bar{4}_6$	4	1667
14	196	2	49	$\bar{4}_6$	8	6403
16	256	8	1	$\bar{4}_6$	1	10923
20	400	4	25	$\bar{4}_6$	4	26667

$y^2 = b \in \bar{4}_6 \Rightarrow b^2 \in \bar{4}_6$, and so $b^2 = 6R_4 + 1$

with R_4 given by:

$$R_4 = 2n_2'(3n_2' + 1) \tag{2.3.8}$$

in which n_2' is the row of *b*. For example, with *y*=19, $n_2' = 60$.

The prime from $a^2 + b^2$ falls in an odd row in $\bar{2}_6$ and

$$\begin{aligned} \bar{1}_6 + \bar{4}_6 &= 6R_1 - 2 + 6R_4 + 1 \\ &= 6R_2 - 1 \end{aligned} \tag{2.3.9}$$

so that $R_1 + R_4 = R_2$. (2.3.10)

Thus, the row of the prime is given by

$$r_p = 2^{2m}\left(6n_2^2 + 2n_2 + \tfrac{1}{6}\right) + \tfrac{1}{3} + 2n_2'(3n_2' + 1) \tag{2.3.11}$$

Table 2.3.4. Some examples for the $(\bar{1}_6\bar{4}_6), (a^2, b^2)$ couples

x	a	m	Q	r_4	R_1
14	196	2	49	8	$2^4(6\times 64 + 2\times 8 + \frac{1}{6}) + \frac{1}{3} = 6403$

y	b	n_2'	R_4 (Eq (2.4)	$R_2 = R_1 + R_4$	Prime $(6R_2 - 1)$
5	25	4	$2\times 4(3\times 4 + 1) = 104$	6507	39041
13	169	28	4760	11163	666977
17	289	48	13920	20323	121937
19	361	60	21720	28123	1688737

Some examples for the $(\bar{1}_6\bar{4}_6), (a^2, b^2)$ couples (⊗ in Table 2.3.1) are given in Table 2.3.4. A similar analysis may be made for the $(\bar{1}_6, \bar{6}_6)$ couple. In this case, R_6 is given by

$$R_6 = 6(q^2 + q) + 1 \tag{2.3.12}$$

with

$$q = 1 + 12\left(\tfrac{1}{2} j(j+1)\right) = 1 + 6r_6(r_6 + 1) \tag{2.3.13}$$

with r_6 the row of y, and, since

$$\begin{aligned}\bar{1}_6 + \bar{6}_6 &= 6R_1 - 2 + 6R_6 + 3 \\ &= 6(R_1 + R_6) + 1 \\ &= 6R_4 + 1\end{aligned} \tag{2.3.14}$$

so that

$$R_1 + R_6 = R_4 \tag{2.3.15}$$

in which R_4 is the row of the prime in $\bar{4}_6$. For example, with x = 16, y = 21, $R_1 = 10923$ (Table 3), and with $r_6 = 3, q = 73$ so that $R_6 = 32413$, hence

$$R_4 = 10923 + 32413 = 43336,$$

so that the prime

$$(6R_4 + 1) = 260017.$$

Note that R_4 is even, as expected. Of course, when calculated directly

$$x^4 + y^4 = 260017,$$

but this result gives no information about the underlying row structure.

Then with r_i, r_j the rows of x, y, respectively

$$x^4 + y^4 = n_4\left(f\left(r_i^4, r_j^4\right)\right) + n_3\left(f\left(r_i^3, r_j^3\right)\right) + n_2\left(f\left(r_i^2, r_j^2\right)\right) + n_1\left(f\left(r_i, r_j\right)\right) + n_0. \tag{2.3.16}$$

The functions for the various (x, y) couples are given in Table 2.3.5. This table is useful for ascertaining factors of $N = x^4 + y^4$. The five terms permit N to be broken up into more easily factored components.

The prime 17 is a common factor and Table 2.3.6 shows the factor structure of the various class couples that yield 17 as a factor.

Although $n_0 = 17$ for $\left(\bar{1}_6, \bar{2}_6\right), \left(\bar{1}_6, \bar{4}_6\right), \left(\bar{5}_6, \bar{4}_6\right), \left(\bar{5}_6, \bar{2}_6\right)$, this does not give them the highest percentage of 17|N (Table 7). Note that when $n_0 = 17, \left(x^4 + y^4\right) \in \bar{2}_6$, so that this sum can never equal a fourth power because this class contains no powers. The remaining class couples $\left(\bar{1}_6, \bar{6}_6\right), \left(\bar{5}_6, \bar{6}_6\right), \left(\bar{3}_6, \bar{2}_6\right), \left(\bar{3}_6, \bar{4}_6\right) \in \bar{4}_6$ and $\left(\bar{3}_6, \bar{6}_6\right) \in \bar{6}_6$ each have $3|x$ and/or $3|y$. Equation (2.3.16) should also be useful in the analysis of the additive components of primes formed from $\left(x^4 + y^4\right)$.

The right end digit (RED) of $\left(x^4 + y^4\right)^* = 1,7$; the asterisk indicates a RED. Since all odd $\left(N^4\right)^* = 1,5$, then $\left(x^4 + y^4\right)^* = 7$ can never equal z^4. Furthermore, even when $\left(x^4 + y^4\right)^* = 1$, if the sum falls in $\bar{2}_6$, then it can never equal z^4, because $\bar{2}_6$ contains no powers. All odd $\left(N^4\right)$ have rows $R_1^* = 0,4, N^4 \in \bar{4}_6$ and $R_6^* = 3,7, N^4 \in \bar{6}_6$. The class couples $\left(\bar{1}_6, \bar{6}_6\right)$ and $\left(\bar{3}_6, \bar{2}_6\right)$, Table 2.3.6, that have $\left(x^4 + y^4\right)^* = 1$, fall in $\bar{4}_6$ which contains even powers. All the (x,y) couples in this case have $5|x$ or $5|y$, as well as $3|x$ or $3|y$. The row, R_4, of the $\left(x_4 + y^4\right)$ sum, if the sum gives an integer to the fourth power, is:

$$R_4 = 8K(12K+1) \tag{2.3.17}$$

with

$$K = \tfrac{1}{2}n(3n+1),$$

n is the row of $\left(x^4+y^4\right)^{\frac{1}{4}}$, or, using the row n_2' of $\left(x^4+y^4\right)^{\frac{1}{2}}$,

$$R_4 = 2n_2'\left(3n_2'+1\right)$$

(Section 2). Of course, from Fermat's Last Theorem $\left(x^4+y^4\right) \neq z^4$, so K can never be an integer. However, one can see that by using integer structure, the analysis can be reduced to a few specific cases where z^4 is possible. Let us consider such cases from Table 6. A composite integer $N \in \overline{4}_6$ is given by:

$$N = p^2 + 6pt, \tag{2.3.18}$$

where p is the lowest prime factor, so that

$$t = \tfrac{1}{6p}\left(N - p^2\right) \tag{2.3.19}$$

As can be seen from Table 2.3.8, $t^* = 0,4,6$. When $N = \left(p_1 p_2 p_3 \ldots\right)^4$ or p_1^4, t always has a factor p_1.

The four sums in Table 2.3.6 that have a RED=1 and fall in $\overline{4}_6$, and are hence potentially equal to an integer power, have values of t incompatible

Table 2.3.5. Components of Equation (2.3.1)

x	y	x^4+y^4	$n_4\left(f\left(r_i^4, r_j^4\right)\right)$	$n_3\left(f\left(r_i^3, r_j^3\right)\right)$	$n_2\left(f\left(r_i^2, r_j^2\right)\right)$	$n_1\left(f\left(r_i, r_j\right)\right)$	n_0
$\bar{1}_6$	$\bar{2}_6$	$\bar{2}_6$	$6^4\left(r_1^4 + r_2^4\right)$	$-4\times 6^3\left(2r_1^3 + r_2^3\right)$	$6^3\left(4r_1^2 + r_2^2\right)$	$-4\times 6\left(8r_1 + r_2\right)$	17
$\bar{1}_6$	$\bar{4}_6$	$\bar{2}_6$	$6^4\left(r_1^4 + r_4^4\right)$	$-4\times 6^3\left(2r_1^3 - r_4^3\right)$	$6^3\left(4r_1^2 + r_4^2\right)$	$-4\times 6\left(8r_1 - r_4\right)$	17
$\bar{1}_6$	$\bar{6}_6$	$\bar{4}_6$	$6^4\left(r_1^4 + r_6^4\right)$	$-4\times 6^3\left(2r_1^3 - 3r_6^3\right)$	$6^3\left(4r_1^2 + 9r_6^2\right)$	$-4\times 6\left(8r_1 - 27r_6\right)$	97
$\bar{5}_6$	$\bar{4}_6$	$\bar{2}_6$	$6^4\left(r_5^4 + r_4^4\right)$	$4\times 6^3\left(2r_5^3 + r_4^3\right)$	$6^3\left(4r_5^2 + r_4^2\right)$	$4\times 6\left(8r_5 + r_4\right)$	17
$\bar{5}_6$	$\bar{2}_6$	$\bar{2}_6$	$6^4\left(r_5^4 + r_2^4\right)$	$4\times 6^3\left(2r_5^3 - r_2^3\right)$	$6^3\left(4r_5^2 + r_2^2\right)$	$4\times 6\left(8r_5 - r_2\right)$	17
$\bar{5}_6$	$\bar{6}_6$	$\bar{4}_6$	$6^4\left(r_5^4 + r_6^4\right)$	$4\times 6^3\left(2r_5^3 + 3r_6^3\right)$	$6^3\left(4r_5^2 + 9r_6^2\right)$	$4\times 6\left(8r_5 + 27r_6\right)$	97
$\bar{3}_6$	$\bar{2}_6$	$\bar{4}_6$	$6^4\left(r_3^4 + r_2^4\right)$	$-4\times 6^3\left(r_2^3\right)$	$6^3\left(r_2^2\right)$	$-4\times 6\left(r_2\right)$	1
$\bar{3}_6$	$\bar{4}_6$	$\bar{4}_6$	$6^4\left(r_3^4 + r_4^4\right)$	$4\times 6^3\left(r_4^3\right)$	$6^3\left(r_4^2\right)$	$4\times 6\left(r_4\right)$	1
$\bar{3}_6$	$\bar{6}_6$	$\bar{6}_6$	$6^4\left(r_3^4 + r_6^4\right)$	$2\times 6^4\left(r_6^3\right)$	$6^3\left(9r_6^2\right)$	$4\times 6\left(27r_6\right)$	81

Table 2.3.6. Factor structure of the various class couples that yield 17 as a factor

x	y	r_i	r_j	x,y classes	Component sums from Table 2.3.5	$x^4 + y^4$	Factors $17\times$
2	13	0	2	$\bar{5}_6\ \bar{4}_6$	$(1+2+3+4)=17\times 1680$ $(5)=17$	2857 ($\bar{2}_6$)	41^2
2	21	0	3	$\bar{5}_6\ \bar{6}_6$	$(1+2+3+4+5)=17^2\times 673$	194497 ($\bar{4}_6$)	17×673
4	9	1	1	$\bar{1}_6\ \bar{6}_6$	$(2+3)=17\times 216$ $(1+4+5)=17\times 185$	6817 ($\bar{4}_6$)	401
4	15	1	2	$\bar{1}_6\ \bar{6}_6$	$(1)=17\times 6^4$ $(2+3+4+5)=17\times 1697$	50881 ($\bar{4}_6$)	41×73
4	19	1	3	$\bar{1}_6\ \bar{4}_6$	$(1+2+3+4)=17\times 7680$ $(5)=17\times 1$	130577 ($\bar{2}_6$)	7681
4	25	1	4	$\bar{1}_6\ \bar{4}_6$	$(1+2+3+4)=17\times 2^4\times 1437$ $(5)=17\times 1$	390881 ($\bar{2}_6$)	22993
6	3	1	0	$\bar{3}_6\ \bar{6}_6$	$(1+5)=1377=17\times 3^4$ $(2+3+4)=0$	1377 ($\bar{6}_6$)	3^4
6	5	1	1	$\bar{3}_6\ \bar{2}_6$	$(1+2+3=4+5)=17\times 113$	1921 ($\bar{4}_6$)	113

Table 2.3.6. (Continued)

x	y	r_i	r_j	x,y classes	Component sums from Table 2.3.5	$x^4 + y^4$	Factors $17\times$
8	1	1	0	$\bar{5}_6\ \bar{4}_6$	$(1+2+3+4)=17\times 240$ $(5)=17\times 1$	4097 ($\bar{2}_6$)	241
8	13	1	2	$\bar{5}_6\ \bar{4}_6$	$(1)=17\times 6^4$ $(2+3+4+5)=17\times 5^4$	32657 ($\bar{2}_6$)	17×113
8	21	1	3	$\bar{5}_6\ \bar{6}_6$	$(3)=17\times 6^3\times 5$ $(1+2+4+5)=17\times 10601$	198577 ($\bar{4}_6$)	11681
10	3	2	0	$\bar{1}_6\ \bar{6}_6$	$(3+5)=17\times 11\times 19$ $(1+2+4)=17\times 3\times 2^7$	10081 ($\bar{4}_6$)	593
12	7	2	1	$\bar{3}_6\ \bar{4}_6$	$(1)=17\times 6^4$ $(2+3+4+5)=17\times 5\times 13$	23137 ($\bar{4}_6$)	1361
12	23	2	4	$\bar{3}_6\ \bar{2}_6$	$(1)=17\times 16\times 6^4$ $(2+3+4+5)=-17\times 5\times 13\times 47$	300577 ($\bar{4}_6$)	17681
12	27	2	4	$\bar{3}_6\ \bar{6}_6$	$(1)=17\times 2^4\times 6^4$ $(2+3+4+5)=17\times 11745$	552177 ($\bar{6}_6$)	$3^4\times 401$

x	y	r_i	r_j	x,y classes	Component sums from Table 2.3.5	$x^4 + y^4$	Factors $17\times$
14	23	2	4	$\bar{5}_6\ \bar{2}_6$	$(1+5)=17\times 20737$ $(2+3+4)=-17\times 2016$	318257 ($\bar{2}_6$)	97×193
14	27	2	4	$\bar{5}_6\ \bar{6}_6$	$(1)=17\times 20736$ $(2+3+4+5)=17\times 12785$	569857 ($\bar{4}_6$)	33521
16	9	2	1	$\bar{1}_6\ \bar{6}_6$	$(2)=-17\times 2^5\times 81$ $(1+3+4+5)=17\times 6833$	72097 ($\bar{4}_6$)	4241
16	15	3	2	$\bar{1}_6\ \bar{6}_6$	$(2+5)=-17\times 7^2\times 31$ $(1+3+4)=17\times 2^5\times 3\times 87$	116161 ($\bar{4}_6$)	6833
16	19	3	3	$\bar{1}_6\ \bar{4}_6$	$(1+2+3+4)=17\times 2^8\times 3^2\times 5$ $(5)=17\times 1$	195857 ($\bar{2}_6$)	41×281
16	25	3	4	$\bar{1}_6\ \bar{4}_6$	$(1+2+3+4)=17\times 26832$ $(5)=17\times 1$	456161 ($\bar{2}_6$)	26833

Table 2.3.7. Percent with factor 17 from Table 2.3.6

Class couples	%with factor 17	n_0
$\bar{1}_6\ \bar{2}_6$	0	17
$\bar{1}_6\ \bar{4}_6$	19	17
$\bar{1}_6\ \bar{6}_6$	24	97
$\bar{3}_6\ \bar{2}_6$	9.5	1
$\bar{3}_6\ \bar{4}_6$	5	1
$\bar{3}_6\ \bar{6}_6$	9.5	81
$\bar{5}_6\ \bar{2}_6$	5	17
$\bar{5}_6\ \bar{4}_6$	14	17
$\bar{5}_6\ \bar{6}_6$	14	97

with N^4, that is, t has factors p' with $(p')^* \neq 7$ (Table 2.3.9); $p^*=7$ is required when $p=17$ for N^4. The same type of analysis may be made for $(x^4 + y^4)$combinations that give factors with REDs 1,3,5,9.

Table 2.3.8. Some factors

N^*	1	3	5	7	9
$(N^4)^*$	1	1	5	1	1
t^* for N^4	0	4	0	6	0
factor p^*	1	3	5	7	9

Table 2.3.9. Factors of *t*

$x^4 + y^4$	t	Factors of t
50881	496	$2^4 \times 31$
1921	16	$2^4 \times 1$
10081	96	$2^5 \times 3$
116161	1136	$2^4 \times 71$

Lists of prime factors of $\left(a^{2^m} + b^{2^m}\right)$ with $(a,b) \in \{(2,3),(2,5),(4,5),(6,5)\}$ in the form $\left(k2^n + 1\right)$ are available (Riesel 1994). In the modular ring Z_4, the primes are $(4r_1 + 1) \in \bar{1}_4$ or $(4r_3 + 3) \in \bar{3}_4$ so that $r_1 = k2^{n-2}$ and $r_3 = \frac{1}{2}\left(k2^{n-1} - 1\right)$, while for Z_6, when $p \in \bar{2}_6, r_2 = \frac{1}{6}\left(k2^n + 2\right)$ and for $p \in \bar{4}_6, r_4 = \frac{1}{6}\left(k2^n\right)$. An integer solution for the row defines the class. For example, with (*a*,*b*)=(2,3), *m*=3 yields

$$\begin{aligned} &= 6817 \\ 2^8 + 3^8 &= 6 \times 1136 + 1 \\ &= 4 \times 1704 + 1, \textit{ for } Z_4, \end{aligned}$$

with *p*=17, *k*=1, *n*=4 so for $Z_4, r_1 = 4, r_3 = \frac{7}{2}$, so $p \in \bar{1}_4$. For Z_6 $r_2 = 3, r_4 = \frac{8}{3}$ so $p \in \bar{2}_6$ with an odd row.

Integer structure (IS) analysis can thus permit a much wider treatment of some numerical systems. In the present case, IS analysis has provided a variety of restraints and functions that illustrate the extreme difficulty of ever finding an integer *z* such that $z^4 = x^4 + y^4$. Furthermore, the ease of producing a prime from $\left(x^4 + y^4\right)$ can often be more readily assessed with class analysis. The analyses here are applicable to all even power sums when $\frac{1}{2}n$ is even. However, when $\frac{1}{2}n$ is odd, the sum effectively becomes an odd-powered sum and can be factored so that there will be no primes. For instance,

$$x^6 + y^6 = \left(x^2\right)^3 + \left(y^2\right)^3, \qquad (2.3.20)$$

and, if x^2 and y^2 are substituted for x and y in Equation (2.3.1) this yields Equation (2.3.4). Similarly,

$$x^{10} + y^{10} = \left(x^2\right)^5 + \left(y^2\right)^5, \qquad (2.3.21)$$

which can be factored according to Equation (2.3.2), and so on.

The overall results, while characterizing the primes, illustrate the impossibility of producing an integer to the fourth power from these sums. The primes, nR, in Table 2.3.4 are all in class $\underline{2}_6$ to see where they fit in the prime sequences in Table 2.3.10.

Some primes from $\{21j+19\}$ are displayed in Table 2.3.11.

Table 2.3.10. Imbedded sequences

p	R	n	n ε	Imbedded sequences
39041	1	3904	{3t +1}	{21j +19}, j = 135
66977	7	6697	{3t + 1}	{21j +19}, j = 318
121937	7	12193	{3t + 1}	{21j +13}, j = 580
1688737	7	168873	{3t}	{21j +12}, j = 8041

Table 2.3.11. Some primes from $\{21j+19\}$

j	0	1	2	3
primes	191,193,197,199	401,409	613,617,619	821,823,827,829
	$\underline{2}_6\ \underline{4}_6\ \underline{2}_6\ \underline{4}_6$	$\underline{2}_6\ \underline{4}_6$	$\underline{4}_6\ \underline{2}_6\ \underline{4}_6$	$\underline{2}_6\ \underline{4}_6\ \underline{2}_6\ \underline{4}_6$
j	4	5	6	7
primes	1031,1033,1039	1249	1451,1453,1459	1663,1667,1669
	$\underline{2}_6\ \underline{4}_6\ \underline{4}_6$	$\underline{4}_6$	$\underline{2}_6\ \underline{4}_6\ \underline{4}_6$	$\underline{4}_6\ \underline{2}_6\ \underline{4}_6$

2.4. PRIME DISTRIBUTIONS IN PRIME ROWS OF THE MODULAR RING Z_4

Modular rings are effectively sorting devices that sort integers into classes within which integers have certain common characteristics. For example, the classes $\bar{1}_4$ and $\bar{3}_4$ in the modular ring Z_4 (Table 2.4.1) contain only odd integers.

Table 2.4.1. Classes and rows for Z4

Row $r_i \downarrow$	Class $i \rightarrow$	$\bar{0}_4$	$\bar{1}_4$	$\bar{2}_4$	$\bar{3}_4$	Comments
0		0	1	2	3	$N = 4r_i + i$
1		4	5	6	7	*even* $\bar{0}_4, \bar{2}_4$
2		8	9	10	11	$(N^n, N^{2n}) \in \bar{0}_4$
3		12	13	14	15	*odd* $\bar{1}_4, \bar{3}_4; N^{2n} \in \bar{1}_4$

However, integers in class $\bar{1}_4$ can equal a sum of squares whereas those in class $\bar{3}_4$ cannot. This follows because classes $\bar{1}_4$ and $\bar{0}_4$ are the only classes with even powers ($\bar{1}_4 = \bar{1}_4 + \bar{0}_4$). Thus all Pythagorean triples have the major component in $\bar{1}_4$; that is, as commonly expressed:

$$x^2 = y^2 + z^2 \text{ with } x = a^2 + b^2, y = 2ab, z = a^2 - b^2.$$

Furthermore, the Golden Ratio family is contained within class $\bar{1}_4$ (Leyendekkers and Shannon 2015).

$$\varphi = \tfrac{1}{2}\left(1 + \sqrt{a}\right) \tag{2.4.1}$$

since all a are found there. Class $\bar{3}_4$ contains no even powers which explains why this class has more primes than $\bar{1}_4$ in regions where even powers are numerous. Other modular rings have their own unique properties (Leyendekkers et al 1997). In this section we examine some of the distribution properties of primes in Z_4 with the use of right-end-digit (RED) properties

Sequences such as the Fibonacci and Mersenne primes depend on a two-prime relationship. A Fibonacci number is a prime only when the sequential index number is a prime. A Mersenne number is a prime when the sequential number is a prime.

In a similar manner we can consider sequences of primes associated with the rows in Table 1. The RED of the prime row p^* will determine the RED of the integer N_p^* in the prime row in class $\bar{1}_4$ (Table 2.4.2).

p^* = 7

When $p^* = 1$, $(4p + 1)^* = 5$, so that no primes are produced. We next try $p^* = 7$ and a range up to 2887 giving a range of values of N_n from 29 to 11549. From the structure and the use of REDs

$$N_n = 40n + 29,\ n \geq 0. \tag{2.4.2}$$

Table 2.4.2. REDs of prime rows p^* versus REDs of integers $N_p^* \varepsilon\, \bar{1}_4$

p^*	1	3	7	9
N_p^*	5	3	9	7

Table 2.4.3. N_n versus row, R

	N_n	R
1	prime	prime
2	prime	composite
3	composite	prime
4	composite	composite

Table 2.4.4. 3|*n*

n^*	N_p (prime)	Adjacent (twin) prime	R_p Prime row	$R_p + 4$ 'Nearby' prime
0	29	31	7	11
3	149	151	37	41
6	269	271	67	71
9	389	---	97	101
12	509	---	127	131
27	1109	---	277	281
30	1229	1231	307	311
48	1949	1951	487	491
57	2309	2311	577	---
72	2909	---	727	---
99	3989	---	997	---
108	4349	---	1087	1091
129	5189	---	1297	1301
132	5309	---	1327	---
156	6269	6271	1567	1571
159	6389	---	1597	1601
177	7109	---	1777	---
198	7949	7951	1987	---
201	8069	---	2017	---
243	9749	---	2437	2441
264	10589	---	2647	---
267	10709	10711	2767	---
276	11069	11071	2767	---
288	11549	11551	2887	---

There are four types of N_n in this sequence (Table 2.4.3).

All four types have primes adjacent to N_n and the rows which identify these,

(i) Primes with prime rows for $p^* = 7$: Twenty four primes are produced with eleven adjacent (twin) to a prime N_P following the numerical map of associated primes from (2.1). There are none adjacent to the rows but half the primes have 'nearby' primes equal to (R_p + 1) (Table 2.4.4).

Table 2.4.5. Primes with composite rows: 3|*n*; triangular numbers are underlined rows

n	N_p	Twin prime	Row Rn	Twin prime
2	109	107	27	29
5	229	227	57	59
8	349	347	87	89
17	709	---	177	179
20	829	827	207	---
26	1069	---	267	269
35	1429	1427	357	359
38	1549	---	387	389
41	1669	1667	417	419
*42	1709	---	427	---
44	1789	1787	447	449
50	2029	2027	507	509
*51	2069	---	517	---
56	2269	2267	567	569
*63	2549	2551	637	---
68	2749	---	687	---
*69	2789	2791	697	---
77	3109	---	777	---
80	3229	---	807	---
*84	3389	3391	847	---
86	3469	3467	867	---
92	3709	---	927	929
*105	4229	4231	1057	---
119	4789	4787	1197	---
122	4909	---	1227	---
*141	5669	---	1417	---
143	5749	---	1437	1439
146	5869	5867	1467	---
154	6189	---	1547	1549
155	6229	---	1557	1559
161	6469	---	1617	---
167	6709	---	1677	---
170	6829	6827	1707	---
*171	6869	6871	1717	---
172	6909	---	1727	---

n	N_p	Twin prime	Row Rn	Twin prime
173	6949	6947	1737	---
176	7069	---	1767	---
*180	7229	---	1807	---
182	7309	7307	1827	---
*183	7349	7347	1837	---
188	7549	7547	1887	1889
191	7669	---	1917	---
194	7789	---	1947	1949
*195	7829	---	1957	---
206	8269	---	2067	2069
209	8389	8387	2097	2099
*210	8429	8431	2107	---
215	8629	8627	2157	---
216	8669	---	2167	---
*225	9029	---	2257	---
227	9109	---	2277	---
233	9349	---	2337	2339
*240	9629	9631	2407	---
245	9829	---	2457	2459
248	9949	---	2487	---
251	10069	10067	2517	---
260	10429	10427	2607	2609
269	10789	---	2697	2699
272	10909	---	2727	2729,2731
*273	10949	---	2737	---
278	11149	---	2787	2789

(ii) Primes with composite rows: Sixty-one primes are generated; 29 of these are twin primes. There are also 23 twin primes associated with the rows (Table 2.4.5).

(ii) Composites with prime rows. Of the 83 composites generated for the 107 prime rows, 53 have 3 as a factor. There are 78 twin primes. The values of *n* range from 10 to 285 and some are triangular numbers (Table 2.4.6).

Table 2.4.6. *n* values for composites with prime rows; *: $3 \nmid N_p$; ‡: triangular numbers

1	25	54*	78*‡	94	127	163*	187	226	247	279*
4	31	55‡	79	96*	130	165*	190‡	228*	250	283
10‡	33*	58	82	97	136‡	166	199	229	255*	285
13	34	60*	85	109	142	169	202	234*	261*	
15*‡	36*‡	61	87*	111*	144*	174*	208	235	265	
16	39*	64	88	118	148	178	213*	237*	268	
19	45*‡	67	90*	121	160	184	220	241	270*	
22	46	75*	93*	123*	162*	186*	223	246*	277	

Table 2.4.7. $p^* = 3$; $N_n = 13 + 40n; 13 \le N_n \le 11533$

Group	N_n	Row $=\lvert n \rvert.3$	Remarks
1	prime	prime	• 25 primes generated from 104 *n* values with (11,13) the only twin primes; • 13 primes adjacent to rows
2	prime	composite	• 65 primes generated from 184 *n* values with 17 twin pairs; • 45 primes adjacent to rows; • factor 3 in 51 rows
3	composite	prime	• 79 composites with prime rows generated from 104 *n* values; • 38 primes adjacent to rows
4	composite	composite	• 119 composites from 184 *n* values; • 97 primes adjacent to rows.

(iv) Composites with composite rows. 117 composites occur with composite rows in the range $0 \le n \le 288$. These composites are associated with 121 primes adjacent to N_n and the rows.

p^* **= 3** The sequences are similar to those for $p^* = 7$ (Table 2.4.7).

The prime rows do not yields many primes compared to the composite rows, and adjacent primes are few. When primes occur in the composite rows and have an adjacent prime they are one of a twin. Adjacent primes are common with composite rows that generate primes so that a high prime density occurs for the associated *n*.

$p^* = 9$

Analogous results are obtained from this RED (Table 2.4.8). Primes may be represented in the form of spectra (Leyendekkers et al. 2009). Similarly, plots of n versus prime number density (primes associated with a particular *n*, including adjacent prime to N_R and the rows) can provide extra data when investigating primality in a particular region. As an example the prime density versus *n* is plotted in Figure 2.4.1 for prime rows $p^* = 9$ (groups 1 and 3). The primes commonly occur when the density is stable over 4 to 6 *n* values or occurs at 3 in an unstable area. We again use the classification in Table 2.4.3. When $p^* = 3$, $N_n^* = 5$, so that only $p^* = 1,7$ and 9 will be considered.

Table 2.4.8. $p^* = 9$; $N_n = 77 + 40n; 77 \le N_n \le 11877; 0 \le n \le 292$

Group	N_n	Row $=\|n+1\|.9$	Remarks
1	prime	prime	• 16 primes generated from 103 values of *n*; • Primes always have $3\|n$; • Total adjacent primes is 10
2	prime	composite	• 72 primes generated from 189 values of *n*; • Adjacent primes yield 17 twin prime pairs; • Total adjacent primes is 67
3	composite	prime	• 87 composites generated from 103 values of *n*; • 31 have $3\|n$; • Total adjacent primes is 85
4	composite	composite	• 117 composites generated from 189 value of *n*; • Total adjacent primes is 113

(a) $p^* = 9$ For this RED in Class $\bar{3}_4$ the basic equation from the structure is

$$N_n = 79 + 40n \tag{2.4.3}$$

which yields N_n from values of n with row = $|n + 1|9$: Table 2.4.9 and Figure 2.4.2.

Class $\bar{3}_4$ has no even powers (Table 2.4.1) so that there is, in effect, more space for primes in regions where these powers are common. This is demonstrated by the relatively low yield of primes in prime rows of Class $\bar{1}_4$. Results for $p^* = 9$ are illustrated in Figure 2.4.2 which is similar to Figure 2.4.1 in terms of prime density but not prime yield.

Table 2.4.9. $p^* = 9$; Class $\bar{3}_4$

Group	N_n	Row $=\|n+1\|.9$	Remarks
1	prime	prime	• 42 primes generated from 102 values of n; • 14 twin primes; • Total adjacent primes is 26
2	prime	composite	• 58 primes generated from 189 values of n; • 25 twin prime pairs; • Total adjacent primes is 53; ($3\|n$ is common)
3	composite	prime	• 60 composites generated from 102 values of n; • Total adjacent primes is 40
4	composite	composite	• 131 composites generated from 189 value of n; • Total adjacent primes is 226 • Majority of composites have $3\|N_n$ unless $3\|n$

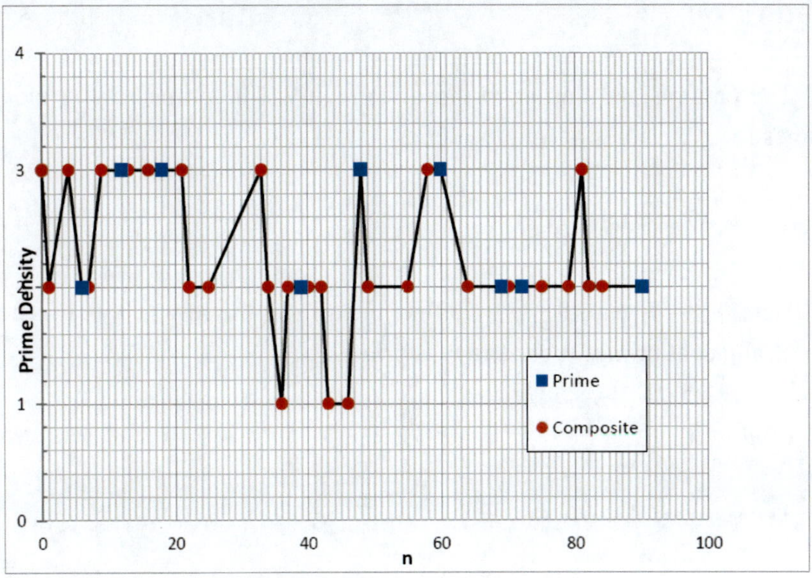

Figure 2.4.1. Class $\underline{1}_4$ $p^* = 9$.

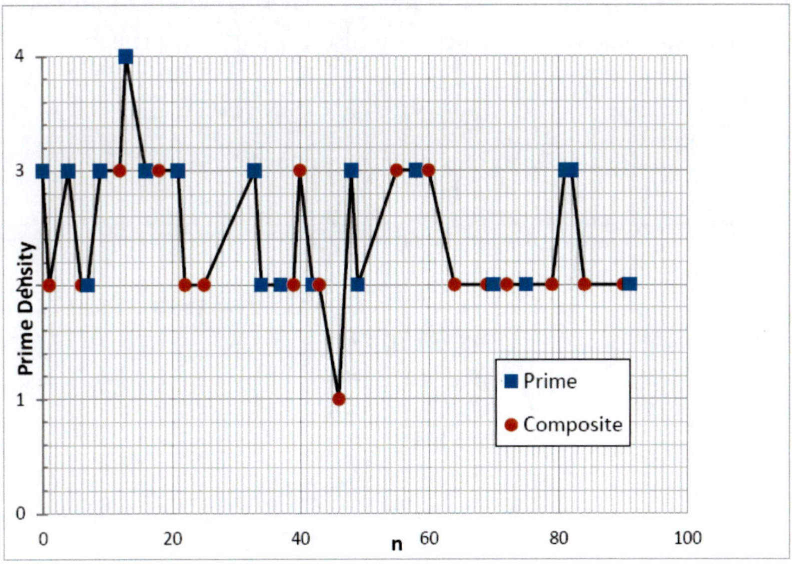

Figure 2.4.2. Class 3_4 $p^* = 9$.

(b) $p^* = 1$

The patterns are similar to those for $p^* = 9$. The prime rows yield 43 primes for 102 values of n in the series

$$N_n = 47 + 40n \tag{2.4.4}$$

Here $N_n^* = 7$, with $0 \le n \le 299$ included here; the row is $|n + 1|$. There are 72 adjacent primes to N_n and r_3 is prime.

(c) $p^* = 7$

The sequence is generated by

$$N_n = 31 + 40n \tag{2.4.5}$$

thus $N_n^* = 1$. Nearly half the 102 values of n yield primes (49 in fact), $0 \le n \le 277$. The row equals $|n|7$. There are also 73 adjacent primes to N_n and rows. The values of n for the primes are displayed in Table 2.4.10.

Table 2.4.10. Values of n for primes, $0 \le n \le 277$

p^*	Row	No. of primes	n values for primes
1	$\|n+1\|1$	43	0,2,3,14,17,23,26,32,39,45,51,56,65,68,87,93,98,102,128,137,144,147,150, 156,173,179,180,186,192,210,212,215,221,224,227,243,261,266,270,285,296,299
7	$\|n\|7$	49	0,1,3,4,6,10,15,22,25,30,36,45,46,48,57,58,64,67,79,87,90, 97,109,118,121,123, 130,136,142,144,156,163,169,174,178,198,201,220,223,228,234,235,237,244,246,265,267,270,276
9	$\|n+1\|9$	42	0,4,7,9,13,16,21,33,34,37,42,48,49,58,70,75,81,82,91,100, 102,111,123,124,126, 130,144,153,165,174,177,187,196,216,229,244,252,264,268,277,280,286

2.5. THE RIGHT-END-DIGIT STRUCTURE OF POWERS IN THE MODULAR RING Z_4

Some of the most challenging conjectures in number theory involve powers of integers in equations and inequalities. As a preliminary to analysing such problems it is useful to consider integers modulo 10, that is, the right-end-digit (RED). We use the elements of Table 2.5.1 to demonstrate the consistency of the RED structure of integer powers within the modular ring Z_4. In theory any modular ring can be used [1], but in practice the fact that of the two odd-integer classes in Z_4 only $\bar{1}_4$ contains even powers. Furthermore, $\bar{0}_4$ contains all the powers of even integers.

Table 2.5.1. Classes and rows for Z_4

Row $r_i\downarrow$	Class $i\rightarrow$	$\bar{0}_4$	$\bar{1}_4$	$\bar{2}_4$	$\bar{3}_4$	Comments
0		0	1	2	3	$N = 4r_i + i$
1		4	5	6	7	*even* $\bar{0}_4, \bar{2}_4$
2		8	9	10	11	$\left(N^n, N^{2n}\right) \in \bar{0}_4$
3		12	13	14	15	*odd* $\bar{1}_4, \bar{3}_4; N^{2n} \in \bar{1}_4$

The power $m \in \bar{0}_4$ is such that

$$m = 4r_0,\ r_0 = 0,1,2,3,4,... \tag{2.5.1}$$

so that with N,M representing the odd and even integers respectively, and an asterisk indicating a RED (Table 2.5.2)

$$\left(N^{4r_0}\right)^* = 1 \tag{2.5.2}$$

or 5 when $N^* = 5$, and

$$\left(M^{4r_0}\right)^* = 6 \tag{2.5.3}$$

or 0 when $M^* = 0$.

Table 2.5.2. $\left(M^{4r_0} \& N^{4r_0}\right)^*$ REDs

M^*	$\left(M^{4r_0}\right)^*$	N^*	$\left(N^{4r_0}\right)^*$
0	0	1	1
2	6	3	1
4	6	5	5
6	6	7	1
8	6	9	1

Example (i): Can $N_1^{4r_0} = N_2^{4r_0} + M^{4r_0}$?

This can be checked by using the RED structure: with $N_1^*, N_2^* \neq 5, M \neq 0$, and $\left(N^{4r_0}\right)^* = 1, \left(M^{4r_0}\right)^* = 6$, then 1≠1 + 6, so the answer is 'no'.

With $N^* = 5$ and $M^* = 0$, the possible combinations (Table 2.5.3) show that two combinations are RED compatible, that is (1,5,6) [i = 4] and (5,5,0) [i = 7].

Table 2.5.3. RED combinations

i	$\left(N_1^{4r_0}\right)^*$	$\left(N_2^{4r_0}\right)^*$	$\left(M^{4r_0}\right)^*$	$\left(N_2^{4r_0}\right)^* + \left(M^{4r_0}\right)^*$	Remarks
1	1	1	6	7	invalid
2	1	5	0	5	invalid
3	5	1	0	1	invalid
4	1	5	6	1	valid
5	5	1	6	7	invalid
6	5	5	6	1	invalid
7	5	5	0	5	valid

Division by 5, $N_1 > N_2$, M and the 5,10 digit structure causes either $N_1^{4r_0}$ > or < the sum so that equality never occurs for the powers in class $\bar{0}_4$, which conforms with Fermat's Last Theorem.

Example (ii): When considering if a particular integer is equal to the sum of integers to a given power it can quickly be assessed which powers will be unacceptable. For odd integers $N_1^* = 1,3,5,7,9$, $(M^{4r_0} + N_2^{4r_0})^* = 1,5,7$, so that 3 and 9 would require more integers with those powers. For even integers, on the other hand, the REDs of the sum of two integers to the power $4r_0$ are 0,2,6, so that $M^* = 4$ or 8 requires more than two, and so on.

When the power $m \in \bar{2}_4$,

$$m = 4r_2 + 2,\ r = 0,1,2,3,..., \tag{2.5.4}$$

so that

$$M^{4r_2+2} = M^2 M^{4r_2}, \tag{2.5.5}$$

$$N^{4r_2+2} = N^2 N^{4r_2} \tag{2.5.6}$$

Results for $4r_0$ yield Table 2.5.4.

Table 2.5.4. REDs for powers for 2 & $4r_2$

M^*	$(M^2)^*$	$(M^2M^{4r_2})^*$	N^*	$(N^2)^*$	$(N^2N^{4r_2})^*$
0	0	0	1	1	1
2	4	4	3	9	9
4	6	6	5	5	5
6	6	6	7	9	9
8	4	4	9	1	1

So $\left(M^2 M^{4r_2}\right)^* = 4$ or 6 except for $M^* = 0$, and $\left(N^2 N^{4r_2}\right)^* = 1$ or 9 except for $N^* = 5$.

Example (i): The sum of an odd and even integer each to the power of $(4r_2 + 2)$ will never equal an odd integer to this power for the couples (1,6) and (9,4) $\left(N_1^2 \neq 3,7\right)$.

Example (ii): When $r_2 = 0$, $m = 2$, so that the powers are squares. The sums and differences of odd and even squares have been analysed previously [2,3,...,8]. However, calculating N_2, M for

$$N_1 = N_2^2 + M^2 \tag{2.5.7}$$

(Clarke et al 1993; Leyendekkers and Shannon 2002 a,b,d; 2007; 2051c; 2016d) is simplified when the RED structure is considered first.

Sums of squares only occur for odd integers when they are in class $\bar{1}_4$. Since classes $\bar{1}_4, \bar{0}_4$ are the only classes with even powers, then

$$\bar{3}_4 \neq \bar{1}_4 + \bar{0}_4. \tag{2.5.8}$$

Tables 2.5.5 and 2.5.6 illustrate the simplification of the calculation of N_2 and M by finding the RED structure first.

Table 2.5.5. $\left(N_2^2 + M^2\right)^*$

	N^*	1	3	5	7	9
	$\left(N^2\right)^*$	1	9	5	9	1
M^*	$\left(M^2\right)^*$					
0	0	1	9	5	9	1
2	4	5	3	9	3	5
4	6	7	5	1	5	7
6	6	7	5	1	5	7
8	4	5	3	9	3	5

Table 2.5.6. REDs for N_2, M

$N_1^* = (N_2^2 + M^2)^*$	(N_2^*, M^*)	A^*
1	(1,0),(5,4),(5,6),(9,0)	1,9,1,9
3	(3,2),(3,8),(7,8)	5,1,5
5	(1,2),(1,8),(3,4),(3,6),(5,0)	3,9,7,9,5
	(7,4),(7,6),(9,2),(9,8)	1,3,1,7
7	(1,4),(1,6),(9,4),(9,6)	5,7,3,5
9	(3,0),(5,2),(5,8),(7,0)	3,7,3,7

Note that when the factors of $N_1 \in \bar{1}_4$ are in class $\bar{3}_4$ there will be no (N_2,M) unless other factors are in class $\bar{1}_4$; for example, $57 \in \bar{1}_4$ but factors (3 and 19) $\in \bar{3}_4$ so no (N_2,M) exists.

Suppose $N = 97$, then from (Leyendekkers and Shannon 2015d;2016d),

$$N_2, M = \tfrac{1}{2}\left(A \pm \sqrt{2 \times 97 - A^2}\right), \tag{2.5.9}$$

and $A = N_2 + M$ with N_2 odd and M even, $A \leq \sqrt{2 \times 97}$. Thus $A \leq 13$. For instance, try 3 in Table 2.5.6 with $N_1^* = 7$; that is, $(N_2^*, M^*) = (9,4)$; this agrees with $97 = 9^2 + 4^2$.

Elements of this class have the form

$$n = 4r_1 + 1,\ r = 0,1,2,3,\ldots\ , \tag{2.5.10}$$

so that

$$N^{4r_1+1} = NN^{4r_1}. \tag{2.5.11}$$

Thus

$$\left(N^{4r_1+1}\right)^* = N^* \tag{2.5.12}$$

and for even integers

$$\left(M^{4r_1+1}\right)^* = (6M)^* = M^*; \tag{2.5.12}$$

Example (i): $6 \times (2,4,6,8,0)^* = (2,4,6,8,0)^*$

Example (ii): With n = 5 in Class $\bar{1}_4$, sums of powers to 5 have been used to yield all the numbers. This flexibility arises from the REDs of the powers being equal to N* or M* so that sums of powers expressed as REDs reduce to addition of simple integers and these sums can have all possible RED values.

Elements of this class have the form

$$n = 4r_3 + 3,\ r = 0,1,2,3,\ldots, \tag{2.5.13}$$

so that

$$N^{4r_3+3} = N^3 N^{4r_3}. \tag{2.5.14}$$

and the REDs of odd integers to a power in $\bar{3}_4$ will be $\left(N^3\right)^*$. Thus for odd integers

$$\left(N^{4r_3+3}\right)^* = \left(N^3\right)^* \tag{2.5.15}$$

and

$$\left(M^{4r_3+3}\right)^* = (6M^3)^* = (M^3)^*; \tag{2.5.16}$$

See Table 2.5.7 for summary.

Table 2.5.7. REDs for $m = 4r_3+3$

N^*	$(N^3)^*$	$(N^{4r_3}N^3)^*$	M^*	$(M^3)^*$	$(M^{4r_3}M^3)^* = (6M^3)^*$
1	1	1	2	8	8
3	7	7	4	4	4
5	5	5	6	6	6
7	3	3	8	2	2
9	9	9	0	0	0

Since all powers in Class $\overline{3}_4$ will have the same REDs as cubes, the RED structures are easily determined even for very large powers in this class so this simplifies the analysis. Of course, the integer structure characteristics assist in the analysis. For example, when $M \in \overline{2}_4, M^{4r_3+3} \in \overline{0}_4$ as there are no powers in $\overline{2}_4$. Thus, $2^7 = 128 = 4 \times 32$, since $(2^3)^* = (2^7)^* = 8$.

In summary:

$$(N^3)^* = (N^{403})^* = (N^{5851})^* = \ldots$$

and

$$(M^3)^* = (M^{5851})^* = (M^{319687})^* = \ldots.$$

We see that a big advantage in considering RED structure is that it is independent of integer size. In a situation dominated by very large integers, the analysis of the properties of powers is simplified. The final RED from a sum of various powers can be readily found and equations can be checked for RED consistency. The results for all powers are summarized in Table 2.5.8.

Table 2.5.8(a). Odd powers

N^*	$\left(N^{4r_1+1}\right)^*$	$\left(N^{4r_3+3}\right)^*$	$\left(N^3\right)^*$	M^*	$\left(M^{4r_1+1}\right)^*$	$\left(M^{4r_3+3}\right)^*$	$\left(M^3\right)^*$
1	1	1	1	0	0	0	0
3	3	7	7	2	2	8	8
5	5	5	5	4	4	4	4
7	7	3	3	6	6	6	6
9	9	9	9	8	8	2	2

Table 2.5.8(b). Even powers

N^*	$\left(N^{4r_0}\right)^*$	$\left(N^{4r_2+2}\right)^*$	$\left(N^2\right)^*$	M^*	$\left(M^{4r_0}\right)$	$\left(M^{4r_2+2}\right)^*$	$\left(M^2\right)^*$
1	1	1	1	0	0	0	0
3	1	9	9	2	6	4	4
5	5	5	5	4	6	6	6
7	1	9	9	6	6	6	6
9	1	1	1	8	6	4	4

It can also be seen that the simplest RED structure occurs when the power is even and in class $\bar{0}_4$ where odd numbers to these powers have a RED of 1 (or 5 when $N^* = 5$) and even numbers to these powers have a RED of 6 (or 0 when $M^* = 0$).

In summary then the primes are in class $\underline{1}_4 \subset Z_4$, a modular ring. The rows of squares in this modular ring are well defined. This permits equalities for the primes to be extracted from the associated integer structure. Furthermore, REDs have the advantage that they are independent of the size of an integer which can be important in the study of some properties of primes.

Chapter 3

FIBONACCI NUMBERS AND PRIMES

If we use the expression $F_p = Kp \pm 1$, p prime, then digital sums of K reveal specific values for primes versus composites in the range $7 \leq p \leq 107$. The associated digital sums of $F_{p\pm1}$ also yield prime/composite specificity. It is shown too that the first digit of F_p, and hence for the corresponding triples, (F_p, $F_{p\pm1}$) and (F_p, F_{p-1}, F_{p-2}) can be significant for primality checks. Sums of the first p Fibonacci numbers, S_p, are shown to be related to K in $F_p = Kp \pm 1$, which is itself a useful indicator of primality for F_p. Digit sums of K, S_p, sums of F_p^2 and Simson's identity are compared. Fibonacci composites with prime subscripts, F_p, have factors $(kp \pm 1)$ in which k is even.

The class of p governs the class of k in the modular ring Z_5, and the digit sum of p, F_p and a function of F_p provide an approximate check on primality. The row structures of the prime-subscripted Fibonacci numbers in the modular ring Z_4 show distinction between primes and composites. The class structure of the Fibonacci numbers suggest that these row structures must survive to infinity and hence that Fibonacci primes must too. The functions $F_p = Kp \pm 1$ and F_p (factors) $= kp \pm 1$ support the structural evidence. The graph of (K/k) versus p displays a Raman-spectra form persisting to infinity: $\ln(K/k)$ is linear in p in the composite case while primes lie along the p-axis to infinity. The structures of Fibonacci numbers, F_n, formed when n equals a prime, p, are analysed using the modular ring Z_5, Pascal's Triangle as well as various properties of the Fibonacci numbers to calculate "Pascal-Fibonacci" numbers to test primality by demonstrating the many structural differences between the cases when F_n is prime or composite.

Sequences of primes $\{P_n\}$ were also obtained by doubling an initial prime and then adding one with this process repeated indefinitely. These sequences have a variety of peculiar structural characteristics. For example, those sequences with initial values of 89 and 137 whose reciprocal values are important, give relatively rich prime yields. Those elements of the sequence with partial values equal to a Mersenne prime can be expressed as a sum of primes.

3.1. Fibonacci Numbers with Prime Subscripts

Rather neat primality checks can be made with $F_p = Kp \pm 1$ and digit sums The structure of a recursive sequence such as the Fibonacci series is, by definition, very regular, so that any fluctuations can be analysed to distinguish between primes and composites when the subscripts or the order in the set are prime numbers.

We have also shown that Fibonacci numbers with prime subscripts equal $Kp \pm 1$, and if composite have factors of the form $kp \pm 1$ (k even) (Leyendekkers and Shannon 2016). Here we continue this analysis. Since $F_p = F_p \times 1$, this function applies generally, as was shown for Fibonacci triples (Table 3.1.1). The K values have digit sums which, like those for other F_p functions can often distinguish between primes and composites (Table 3.1.2).

The right-end-digit (RED) for K, designated by K^*, has distinct values for a given p^*, irrespective of primality (Table 3.1.3).

Table 3.1.1. Digit sums of *K* (*p*: prime; *c*: composite)
‡ ['+' ≡ $p^* \in \{1, 9\}$; '–' ≡ $p^* \in \{3, 7\}$]

p	F_p	K	Sign‡	Type
7	13	2	–	p
11	89	8	+	p
13	233	18	–	p
17	1597	94	–	p
19	4181	220	+	c
23	28657	1246	–	p
29	514229	17732	+	p
31	1346269	43428	+	c
37	24157817	652914	–	c
41	165580141	4038540	+	c
43	433494437	10081266	–	p
47	2971215073	63217342	–	p
53	53316291173	100596778	–	c
59	956722026041	1621562750	+	c
61	2504730781961	41061160360	+	c
67	44945570212853	670829406162	–	c
71	308061521170129	4338894664368	+	c
73	806515533049393	11048157986978	–	c
79	14472334024676221	183194101578180	+	c
83	99194853094755497	1195118711985006	–	p
89	1779979416004714189	19999768719154092	+	c
97	83621143489848422977	862073644225241474	–	c
101	573147844013817084101	5674731128849674100	+	c
103	1500520536206896083277	14568160545698020226	–	c
107	10284720757613717413913	96118885585174929102	–	c

The distribution of these is displayed in Table 3.1.4. A comparison of primes with composites (Table 3.1.5) illustrates that no distinction exists for the last digit. However, the first digit displays distinctions except when p = 7 with 2 as a common digit.

The first digit of each Fibonacci number occurs at a specific position, n, in the series. The following positions occur in a regular pattern as shown by $(n_j - n_{j-1})$ in Table 3.1.6. As noted above, the first digit of F_p seems to offer a distinction between primes and composites.

Table 3.1.2. Digit sums of K

p^*	Primes	Composites
1	8	1,2,3,6,9
3	2, 4, 6,9	2,7,8
7	1, 2, 4	6, 8,9
9	2	3, 4, 6, 8

Table 3.1.3. REDs

p*	K*
1	0+, 8+
3	6–, 8–
7	2–, 4–
9	0+, 2+

Table 3.1.4. First and last digits of F_p

1st digit → / Last ↓	1	2	3	4	5	6	7	8	9
1	✓	✓	✓	✓	✓	✓	✓	×	✓
2	✓	✓	✓	✓	✓	✓	×	✓	×
3	✓	✓	×	✓	✓	✓	✓	✓	×
4	✓	✓	✓	✓	×	×	✓	✓	×
5	✓	✓	✓	×	✓	✓	✓	×	✓
6	✓	✓	✓	✓	×	✓	×	×	✓
7	✓	✓	✓	✓	✓	×	×	×	✓
8	✓	✓	✓	✓	×	✓	×	×	×
9	✓	✓	✓	✓	✓	×	✓	✓	×
0	✓	✓	✓	×	✓	✓	×	✓	×

Table 3.1.5. Comparison of primes and composites

p^*	Primes		Composites	
	1st digit	Last digit	1st digit	Last digit
1	8	9	1,2,3,5	1, 9
3	2, 4,9	3, 7	5	3,7
7	1, 2	3, 7	1,2,4,8	3, 7
9	5	9	1, 4, 9	1, 9

Table 3.1.6. Some patterns

1st digit	1st n_j	$(n_j - n_{j-1})$ patterns
1	7	5, 5, 4, 1, 4, 1, 4, 5, 4, 1, 4, 1, 4, 5, 5, 4, 1, 4, 1, 4, 5, …
2	8	5, 5, 5, 9, 5, 5, 5, 5, 5, 9, 5, 5, 5, 9, …
3	4	5, 5, 1, 4, 5, 5, 1, 4, 5, 1, 4, 5, 5, 1, 4, 5, 5, …
4	19	5, 1, 9, 5, 14, 5, 5, 14, 5, 19, 5, …
5	5	5, 19, 5, 19, 5, 19, 19, 5, …
6	15	5, 19, 24, 19, 5, 19, …
7	25	19, 5, 19, 24, 19, …
8	6	5, 19, 24, 19, 5, …
9	16	43, 24, 19, …

Table 3.1.7. Examples of primality for $n \in \{3t+2\}$

p	F_p	n	Sum of digits of n	R	primality
11	89	8	8	9	p
13	233	23	5	3	p
29	514229	51422	5	9	p
53	etc	5331629117	2	3	c
67		4494557021285	5	3	c
107		1028472075761371741391	5	3	c

Since only $n = p$ yields primes these can be sieved out (Erdös and Jabotinsky 1958). As we see in Chapter Six, the elements of the prime sequences have the form nR where R represent the RED and n represents the digits to the left of R; for example, F_p = =1597 so that $R = 7$ and $n =$ 159. $n \in \{3t\} \vee \{3t+1\} \vee \{3t+2\}$. When $n \in \{3t+2\}$, only $R = 3$ or 9 will yield primes; for instance, when for $p = 13$, $F_p = 233$, $n = 23 = 3 \times 7 + 2$ (Table 3.1.7).

The sum of th digits of n is commonly 5 which renders this quantity indiscriminate for composite testing.

Further analysis along these lines can be made so that indications of primality build up and increase the probability of testing the primality of F_p. The results outlined here can then be extended to consider probabilistic primality testing (Watkins 2014).

3.2. Fibonacci Number Sums as Prime Indicators

Sums of the first p Fibonacci numbers, S_p, are related to K in $F_p = Kp \pm 1$, which itself can be an indicator of primality for F_p as we saw in the last section. The sum of the first n Fibonacci numbers is well-known to be

$$\sum_{i=1}^{n} F_n = F_{n+2} - 1 \tag{3.2.1}$$

which, when $n = p$, becomes

$$S_p = F_{p+2} - 1. \tag{3.2.2}$$

We compare this sum for the primes $3 \le p \le 113$ using the method of digit sums (Grabner et al 1997; Leyendekkers and Shannon 2014). The digit sum in base b, of the digits of the number n, is often represented by $s_b(n)$ (Fujiwara 2005), and satisfies the congruence:

$$n \equiv s_b(n) \pmod{b-1}. \tag{3.2.3}$$

When $b = 10$, this congruence is the basis of high school techniques such as *'casting out nines'* and of *divisibility tests* such as those for 3 and 9.

Table 3.2.1. Digit sums of first *p* Fibonacci numbers and parity

p	Sp	Digit sum	Parity	p	Sp	Digit sum	Parity
3	4	4	p	59	2504730781960	7	c
7	33	6	p	61	6557470319841	8	C
11	332	7	p	67	117669030460993	1	C
13	609	6	p	71	806515533049392	9	C
17	4180	4	p	73	2111485077978049	1	c
19	10945	1	c	79	37889062373143905	6	c
23	75024	9	p	83	259695496911122584	7	p
29	1346268	3	p	89	4660046610375530308	4	c
31	3524577	6	c	97	218922995834555169025	1	c
37	63245985	6	c	101	1500520536206896083276	3	c
41	433494436	4	c	103	3928413764606871165729	6	c
43	1134903169	1	p	107	26925748508234281076008	7	c
47	7778742048	9	p	109	70492524767089125814113	6	c
53	139583862444	3	c	113	483162952612010163284884	4	p

Table 3.2.2. Digit sum for first *p* Fibonacci Numbers: [*p = right-end-digit]**

*p**	Primes	Composites
1	7	3,4,6,8,9
3	1,4,6,7,9	1,3
7	4,6,9	1,6,7
9	3	1,4,6,7

Table 3.2.3. Divisibility of neighbours of F_p by *p*

*p**	F_{p+1}	F_{p-1}
1		✓
3	✓	
7	✓	
9		✓

The sums of Fibonacci numbers and the corresponding digit sums (Tables 3.2.1 and 3.2.2) show distinctions between primes and composites (Table 3.2.2).F_{97}, if a prime, does not conform so that it was checked with the factor ($k\backslash Kp \pm 1$) technique which showed that it was indeed a composite, namely

$$F_{97} = 83621143489848422977$$
$$= 193 \times 389 \times 1113805073322701$$

with $193 = 2 \times 97 - 1$

and $389 = 4 \times 97 + 1.$

However, either F_7 or F_{37} do not conform, both having a digit sum of 6. F_{109}, if conforming, is a composite, and F_{113} is a prime. These results are generally in accord with the K results, from Leyendekkers and Shannon (2014)

$$F_p = Kp \pm 1 \tag{3.2.4}$$

The K values appear to be more reliable.

One of the direct neighbours of F_p is divisible by p according to p^* in a neat pattern (Table 3.2.3).

$p^* = 3, 7$

From (3.2.2) and the Fibonacci recurrence relation

$$S_p + 1 = F_{p+1} + Kp - 1 \tag{3.2.5}$$

$$F_{p+2} = F_{p+1} + F_p \tag{3.2.6}$$

so that

$$K = \frac{1}{p}\left(S_p + 2 - F_{p+1}\right). \tag{3.2.7}$$

Both (S_p+2) and F_{p+1} are divisible by p.

p* = 1, 9

Since from repeating the Fibonacci recurrence relation

$$\begin{aligned} F_{p+2} &= 2F_p + F_{p-1} \\ &= 2(Kp+1) + F_{p-1} \end{aligned} \tag{3.2.8}$$

and

$$\begin{aligned} S_p &= 2(Kp+1) + F_{p-1} - 1 \\ &= 2Kp + 1 + F_{p-1} \end{aligned} \tag{3.2.9}$$

then

$$K = \frac{1}{2p}\left(S_p - 1 - F_{p-1}\right) \tag{3.2.10}$$

Table 3.2.4. *K* values from Equation (3.2.3)

p	Parity	K	p	Parity	K
3	p	1	53	c	1005967758
7	p	2	67	c	670829406162
13	p	18	73	c	11048157986978
17	p	94	83	p	1195118711985006
23	p	1246	97	c	862073644225241474
37	c	652914	103	c	14568160545698020226
43	p	10081266	107	c	96118885585174929102
47	p	63217342	113	p	16332019981684334481118

Table 3.2.5. *K* values from Equation (3.2.6)

p	Parity	K	p	Parity	K
11	p	8	61	c	41061160360
19	c	220	71	c	4338894664368
29	p	17732	79	c	183194101578180
31	c	43428	89	c	19999768719154092
41	c	4038540	101	c	5674731128849674100
59	c	16215627560	109	c	494050431343748276624

Table 3.2.6. Digit sums of *K*

p^*	Primes	Composites
1	8	3,6
3	1,4,6,9	3
7	1,2,4	6,8,9
9	2	3,4,5,6

Table 3.2.7. Digit sums of Fibonacci squares

p	Parity	F_p	F_{p+1}	F_pF_{p+1}	p	Parity	F_p	F_{p+1}	F_pF_{p+1}
3	p	2	3	6	59	c	8	9	9
7	p	4	3	3	61	c	8	8	1
11	p	8	9	9	67	c	5	6	3
13	p	8	8	1	71	c	1	9	9
17	p	4	1	4	73	c	1	1	1
19	c	5	6	3	79	c	4	3	3
23	p	1	9	9	83	p	8	9	9
29	p	5	8	4	89	c	4	1	4
31	c	4	3	3	97	c	1	1	1
37	c	8	8	1	101	c	5	8	4
41	c	4	1	4	103	c	4	5	2
43	p	5	6	3	107	c	8	8	1
47	p	1	9	9	109	c	8	8	1
53	c	5	8	4	113	p	4	1	4

Table 3.2.8. Digit sums

p*	Primes	Composites
1	9	1,3,4,9
3	1,2,3,4,6,9	1,2,4
7	3,4,9	1,3
9	4	1,3,4,9

As for Section (3.1), K values agree with previous values. The sums of digits of K from Tables 3.2.4 and 3.2.5 show clear distinctions for primes and composites (Table 3.2.6).

The digit sums of Fibonacci squares were calculated from

$$F_p \times F_{p+1} = \sum_{i=1}^{p} F_p^2. \qquad (3.2.11)$$

The digit sums generally show distinctions between primes and composites (Tables 3.2.7, 3.2.8) but not as reliably as K or S_p. The digit sums are simply obtained from the digit sums of F_p times the digit sums of F_{p+1}. The digit sums 3 and 4 are clearly indeterminate for this index. An approximation to this identity leads to the Golden Ratio, namely

$$F_{n+1}F_{n-1} \approx F_n^2$$

whereas the precise form is

$$F_{n+1}F_{n-1} = F_n^2 + (-1)^n \qquad (3.2.12)$$

and $F_{n+2}F_{n-2} = F_n^2 + (-1)^{n-1}$ (3.2.13)

and the digit sums of these quantities show that the digit sum of F_p^2 is either 1 or 7. (Compare the perfect numbers which always have a digit sum of 1.)

While the results are interesting because of the links, other indicators such as K or S_p are more useful for primality tests. Furthermore, digit sums provide patterns for further exploration in both pure (Shallit 1985) and applied mathematics (Grabner et al 1997) for university student projects at all levels, as do right-end-digits as integers (modulo 10) (Watkins, 2014). In particular, such analyses have not yet been done for the generalized Fibonacci numbers of Chapter 4.

Table 3.2.9. Digit sums for F_p^2

p	Parity of F_p	F_{p-1}x F_{p+1}	F_{p-2}x F_{p+2}	F_p^2	p	Parity of F_p	F_{p-1}x F_{p+1}	F_{p-2}x F_{p+2}	F_p^2
11	p	9	2	1	7	p	6	8	7
31	c	6	8	7	17	p	6	8	7
41	c	6	8	7	37	c	9	2	1
61	c	9	2	1	47	p	9	2	1
71	c	9	2	1	67	c	6	8	7
101	C	6	8	7	97	c	9	2	1
13	p	9	2	1	107	c	9	2	1
23	p	9	2	1					
43	p	6	8	7	19	c	6	8	7
53	c	6	8	7	29	p	6	8	7
73	c	9	2	1	59	c	9	2	1
83	P	9	2	1	79	c	6	8	7
103	C	6	8	7	89	c	6	8	7
113	p	6	8	7	109	c	9	2	1

3.3. FIBONACCI PRIMES[4]

In this section we continue the ideas of the previous two sections by providing a simple method to determine the factors for Fibonacci composites with prime subscripts, F_p. In [3, 4] it was shown that many F_p

[4] Based on Leyendekkers and Shannon (2014).

properties are closely related to the modular ring Z_5 (Table 3.3.1) and we find that here too.

Table 3.3.1. The modular ring Z_5

Class	$\bar{0}_5$	$\bar{1}_5$	$\bar{2}_5$	$\bar{3}_5$	$\bar{4}_5$
Row	$5r_0$	$5r_1+1$	$5r_2+2$	$5r_3+3$	$5r_4+4$
0	0	1	2	3	4
1	5	6	7	8	9
2	10	11	12	13	14
3	15	16	17	18	19
4	20	21	22	23	24
5	25	26	27	28	29

The factors were found to have the form ($kp \pm 1$) in which k is even (Table 3.3.2) and in a specific class determined by p (Table 3.3.3).

Table 3.3.2. Composite Fibonacci numbers

p	F_p	k	Factors kp ± 1	sign
19	4181	2, 6	37, 113	− −
31	1346269	18, 78	557, 2417	− −
37	24157817	2, 4, 60	73, 149, 2221	− + +
41	165580141	68, 1448	2789, 59369	+ +
53	53316291173	18, 1055580	953, 55945741	− +
59	956722026041	6, 4593662	353, 2710260697	− −
61	2504730781961	74, 9098418	4513,555003497	− −
67	44945570212853	4, 2493789614	269,167083904137	+ −
71	308061521170129	94, 659216494	6673,46165371073	− −
73	806515533049393	128436, 1178366	9375829, 86020717	+ −
79	14472334024676221	2, 1166841411326	157, 92180471494753	− −
89	17799794160047 14189	12, 187088575 48320	1069, 16650883218 00481	+ +
101	57314784401381708 4101	7361578, 7632257214	743519377,7708579 78613	− −
103	1500520536206896 083277	5040, 54798, 4972036	51921, 5644193, 512119709	−+ −

Table 3.3.3. Z_5 classes

p*	k*	Class in Z_5
1	4,8	$\overline{4}_5\ \overline{3}_5$
3	6,8,0	$\overline{1}_5\ \overline{3}_5\ \overline{0}_5$
7	2,4,0	$\overline{2}_5\ \overline{4}_5\ \overline{0}_5$
9	2,6,0	$\overline{1}_5\ \overline{2}_5\ \overline{0}_5$

The sums of digits of integers, also known as the digital root function, $\varphi(n)$, following (Atanassov and Shannon 2011; Gardiner 1982; Shannon and Horadam 1991) is given by

$$n = \sum_{i=1}^{k} a_i.10^{k-i} \equiv \overline{a_1 a_2 ... a_k}$$

where a_i is a natural number and $0 \le a_i \le 9$ $(1 \le i \le k)$. Then

$$\varphi(n) = \begin{cases} 0, & n = 0, \\ \sum_{i=1}^{k} a_i, & n > 0. \end{cases}$$

although from this point on in this section it is sufficient to use the decimal count system. Now let us consider the sums of digits of F_p. These give different values for primes and composites except for right-end-digits $p^* = 3$ or 9 where a sum of 5 is non-distinctive. However, these sums give a rough guide to primality or the possibility of F_p being composite (Table 3.3.4).

Table 3.3.4. Sums of digits

p*	Primes sum*	Composites sum*
1	2	1,4,5,8
3	1,5,8	1,5
7	1,4	5,8
9	5	4,5,8

Table 3.3.5. Sums of digits

*p**	Primes sum*	Composites sum*
1	2	6,9
3	3	6,9
7	3,9 (6)	6,9
9	3	3,6,9

Table 3.3.6. Sums of digits of *p*

p*	Sum of digits of p	
	F_p prime	F_p composites
1	2	4,5,7,8
3	2,3,4,5,7	1,4,8
7	2,7,8	1,4,7,8
9	2	1,5,7,8

Another check is to subtract the digital root function $\varphi(F_p)$ from F_p, then divide that number by 3 and sum the digits; that is:

$$\varphi\left(\tfrac{1}{3}\left(F_p - \varphi(F_p)\right)\right)$$

as in Table 3.3.5. Generally, composites have a sum of 6, with F_{47} an exception. Even the sum of digits of *p* show distinctions between primes and composites (Table 3.3.6).

The first factor of the composites generally has a small value for *k* so that primality is relatively easy to determine. The digit sums also provide guides to primality in special sequences such as the Fibonacci numbers

(Ribenboim 2000). The digit sum of an integer follows simply from the integer structure, so we can use this sum to characterise certain identities. For example, perfect numbers, N_p, are represented by $2^{p-1}(2^p - 1)$ where p is a prime (Table 3.3.7).

Table 3.3.7. Digits sums of some perfect numbers

p	N_p	Class of N_p in Z_5	φ (N_p)
2	6	$\bar{1}_5$	6
3	28	$\bar{3}_5$	1
5	496	$\bar{1}_5$	1
7	8128	$\bar{3}_5$	1
11	2096128	$\bar{3}_5$	1
13	33550336	$\bar{1}_5$	1
17	8589869056	$\bar{1}_5$	1

This illustrates $N_p \in \{\bar{1}_5, \bar{3}_5\} \Rightarrow N_p^* \in \{6,8\}$ and $\varphi\{N_p\} = 1, p > 2$. For the Fibonacci primes, another digit sum which can be used comes from a modification of Simson's identity:

$$F_{p-1}F_{p+1} = (F_p - 1)(F_p + 1).$$

The corresponding right-end-digit results are

$$p^* = \begin{cases} 1,9 \ \Rightarrow \dfrac{F_p - 1}{p} \in Z, \\ 3,7 \ \Rightarrow \dfrac{F_p + 1}{p} \in Z, \end{cases}$$

as in Table 3.3.8.

Table 3.3.8. Fibonacci primes digit sums

p*	Sum of digits	
	F_p prime	F_p composites
1	8	1,3, 6,9
3	2, 4, 6, 9	3
7	1, 4	6, 9
9	2	3, 4, 5, 6

The factors and *k* can be more easily calculated if we have data on the RED structure (Table 3..3.9).

Table 3.3.9. Factors and RED structure

p*	k*	Factors*
1	(4,8)(4,4)(8,8)	3,7,9
3	(6,6)(0,8)(0,6,8)	1,7,9
7	(4,4)(0,2,4)	1,3,7,9
9	(0,2,6)	1,3,7,9

3.4. An Infinite Primality Conjecture for Prime-Subscripted Fibonacci Numbers

On knowing the infinite, Franklin (2014) has this to say: "It is evident that the idea of an infinite structure cannot be derived purely from perceptual experience [...] Our perceptual experience is finite in character". This section explores an aspect of infinity in the context of Fibonacci primes and the regularity of the Fibonacci numbers generated from the second order homogeneous recurrence relation

$$F_n = F_{n-1} + F_{n-2},\ n > 2, \qquad (3.4.1)$$

which gives rise to the many periodicities found in this sequence (Livio 2002) and its very precise integer structure; for example, in the associated modular ring Z_4 (Table 3.4.1).

This particular ring is the most appropriate in this context because the prime-subscripted Fibonacci numbers satisfy (Drobot 2000; Somer 2002).

$$F_p = F^2_{\frac{p+1}{2}} + F^2_{\frac{p-1}{2}} \tag{3.4.2}$$

and the only odd class in Z_4, which can form this sum of squares is $\bar{1}_4$, which is generated by $4r_1 + 1$; that is, prime-subscripted Fibonacci numbers will always fall in this class. When F_p is itself prime, Equation (3.4.2) is the only possible sum, but not with composites (Leyendekkers and Shannon 1998; 2002).

The prime-subscripted Fibonacci numbers will now be considered in detail in order to assess the evidence of infinitely many Fibonacci primes. Since there are infinitely many primes, F_p will have infinitely many values, but since F_p may be composite, the number of prime-subscripted Fibonacci primes may be finite. The class structure for the Fibonacci numbers in the modular ring Z_4 is.

Table 3.4.1. Classes and rows for Z_4

Row $r_i \downarrow$	Class $i \rightarrow$	$\bar{0}_4$	$\bar{1}_4$	$\bar{2}_4$	$\bar{3}_4$	Comments
0		0	1	2	3	$N = 4r_i + i$
1		4	5	6	7	even $\bar{0}_4$, $\bar{2}_4$
2		8	9	10	11	$(N^n, N^{2n}) \in \bar{0}_4$
3		12	13	14	15	odd $\bar{1}_4, \bar{3}_4$; $N^{2n} \in \bar{1}_4$

Table 3.4.2. Row structures of F_p

p*	Primes	Composites
1	$\bar{2}_4\,\bar{1}_4\,\bar{1}_4$	$\bar{3}_4\,\bar{1}_4\,\bar{3}_4,\bar{1}_4\,\bar{3}_4\,\bar{2}_4,\bar{1}_4\,\bar{1}_4\,\bar{3}_4,\bar{1}_4\,\bar{0}_4\,\bar{1}_4,$ $\bar{1}_4\,\bar{0}_4\,\bar{3}_4$
3	$\bar{2}_4\,\bar{2}_4\,\bar{2}_4,\bar{0}_4\,\bar{3}_4\,\bar{3}_4,$ $\bar{1}_4\,\bar{2}_4\,\bar{2}_4,\bar{2}_4\,\bar{2}_4\,\bar{2}_4$	$\bar{1}_4\,\bar{3}_4\,\bar{1}_4,\bar{0}_4\,\bar{3}_4\,\bar{0}_4,\bar{1}_4\,\bar{0}_4\,\bar{3}_4$
7	$\bar{3}_4\,\bar{3}_4\,\bar{0}_4,\bar{1}_4\,\bar{0}_4\,\bar{2}_4$	$\bar{2}_4\,\bar{3}_4\,\bar{1}_4,\bar{0}_4\,\bar{0}_4\,\bar{2}_4,\bar{1}_4\,\bar{0}_4\,\bar{0}_4,\bar{2}_4\,\bar{1}_4\,\bar{0}_4$
9	$\bar{1}_4\,\bar{3}_4\,\bar{2}_4$	$\bar{2}_4\,\bar{3}_4\,\bar{0}_4,\bar{3}_4\,\bar{3}_4\,\bar{2}_4,\bar{0}_4\,\bar{3}_4\,\bar{2}_4,\bar{2}_4\,\bar{3}_4\,\bar{2}_4$

$$\bar{1}_4\bar{1}_4\bar{2}_4\bar{3}_4\bar{1}_4\bar{0}_4\bar{1}_4\bar{1}_4\bar{2}_4\bar{3}_4\bar{1}_4\bar{0}_4\bar{1}_4\bar{1}_4\bar{2}_4\bar{3}_4\bar{1}_4\bar{0}_4\ldots \qquad (3.4.3)$$

which is repeated to infinity as the formation of a recursive sequence does not change (Knopp 1990). Since $F_p \in \bar{1}_4$, the row structure is given by

$$F_p = 4r_1 + 1. \qquad (3.4.4)$$

The row structures of F_p for $p = 7$ to 101 are displayed in Table 3.4.2 according to p^*, the right-end-digit of p (that is, $p \pmod{10}$).

This p^* defines the class of p in the modular ring Z_5 (Ter 1996). As can be seen there is a distinction between primes and composites which is worth exploring further. If there are no primes for very large p, then the row structure of F_p would be very restricted for $p^* = 1$ or 3, and some row structures for $p^* = 7$ or 9 would not occur. In view of the precise nature of the F_p structure this would not be possible. This makes a compelling case for infinitely many prime Fibonacci numbers.

We also consider functions and factors defined respectively by $F_p = pK \pm 1$ and F_p (factors) $= pk \pm 1$. When K/k for p 7 to 107 is plotted as function of p, the result is a Raman-like spectra (Lilley 1973) with a base of unity (since $K = k$ for primes), and a variety of bands when $K/k \neq 1$ (representing the composites). Continuous bands without the base of 1 would have to occur if no more primes occur for large p. This would be inconsistent with any normal spectra and would indicate severe rupture of the F_p-sequence structure which would not be possible in view of the formation mechanism of the Fibonacci recurrence relation.

$\ln(K/k)$ as a function of p: for composite F_p this function is linear and passes through the origin where $K = k$ and $\ln(K/k) = 0$; that is, for composites only

$$\ln\left(K/k\right) \approx \frac{p}{10(\varphi - 1)} \tag{3.4.5}$$

in which φ is the Golden Ratio.

Table 3.4.3. Composite F_p

p	$\ln(K/k)^{\ddagger}$	$p/10(\varphi\text{-}1)^{\ddagger}$
19	0	0
31	0.1	0.1
37	0.2	0.1
41	0.2	0.3
53	0.4	0.4
59	0.5	0.5
61	0.5	0.5
67	0.6	0.6
71	0.7	0.7
73	0.7	0.7
79	0.8	0.8
89	0.9	0.9
97	1	1

In Table 3.4.3 the standardized values represented by $\ln(K/k)^{\ddagger}$ and $p/10(\varphi\text{-}1)^{\ddagger}$, respectively are set out to show that the two sides of (3.4.5) are proportionally approximate. The following standard normalization formula was utilized:

$$x_i^{\ddagger} = \frac{x_i - x_{\min}}{x_{\max} - x_{\min}} \tag{3.4.6}$$

When F_p is prime the line coincides with the p-axis which meets the composite line at the origin and increases asymptotically. Another parameter which supports this structural evidence is S_p (Hoggatt 1969):

$$S_p = \sum_{i=1}^{p} F_i = F_{p+2} - 1 \tag{3.4.7}$$

from which, S_p^*, the right-end-digits for S_p, for primes are distinct from those of composite F_p (Table 3.4.4).

The stability of S_p* is based on the structural ability of the Fibonacci numbers from their very definition by means of a linear recurrence relation (3.4.1) and the periodicity of their class structure in the modular ring.

With integers in the form nR (as in the previous sections; for example, F_{29}= 514229, so n =51422 and R = 9), three sequences arise $n \in \{3t\} \vee \{3t+1\} \vee \{3t+2\}$, and so primes and composites can be compared on the basis of their distributions in these three sequences (Table 3.4.5).

Table 3.4.4. S_p, $3 \leq p \leq 97$

p*	S_p^*	
	Primes	Composites
1	2	1,6, 8
3	5	4,9
7	1, 4, 8	5
9	9	6

Table 3.4.5. $n \in \{3t\} \vee \{3t+1\} \vee \{3t+2\}$

n ε	Primes			Composites		
	p	R	n*	p	R	n*
$\{3t\}$	17	7	9	41	1	4
	23	7	5	79	1	2
				97	7	7
				101	1	1
				103	7	7
$\{3t+1\}$	7	3	1	19	1	8
	43	7	3	31	9	6
	47	3	7	37	7	1
	83	7	9	59	1	4
				61	1	6
				71	9	2
				73	3	9
				89	9	8
$\{3t+2\}$	11	9	8	53	3	7
	13	3	3	67	3	5
	29	9	2			

This class pattern of F_p is invariant since the mechanism of generation of the Fibonacci numbers remains the same to infinity. If class structure is invariant, then the demonstrated difference between the row structures for primes and composites should also be invariant. The relationships of K and k with p also show that primes are generated as long as primes exist; that is, since there is an infinity of primes (Watkins 2014), then there must also be an infinity of prime-subscripted Fibonacci prime numbers. Another possible line of research related to the central issue in this section would be by means of asymptotic proofs; that is, for 'almost all n'. These have been used previously for Fibonacci numbers by Horadam (1966) and Subba Rao (1954; 1959).

3.5 PRIMES WITHIN GENERALIZED FIBONACCI SEQUENCES

The analysis of the structure of the Golden Ratio family $\{\varphi_a\}$ can generate sets of generalized Fibonacci sequences (Anatriello and Vincenzi 2014; Whitford 1977). Following on from a study of Fibonacci primes of the first member ($a = 5$) of this 'family,' obvious questions are:

- Do the generalized Fibonacci primes thus generated have properties similar to those of the ordinary Fibonacci sequence?
- Do the primality tests for the ordinary Fibonacci sequence apply more generally?

We have previously (2015) shown that the modular-ring structure is critical in the formation of the Golden Ratio family of generalized Fibonacci sequences, which are generated by

$$F_{\varphi_a}(n+1) = F_{\varphi_a}(n) + r_1 F_{\varphi_a}(n-1),\ n > 1, \qquad (3.5.1)$$

where

$$\varphi_a = \frac{1+\sqrt{a}}{2} \qquad (3.5.2)$$

and with initial conditions $F_{\varphi_a}(1) = F_{\varphi_a}(2) = 1$, and where r_1 is the row of $a \in \bar{1}_4 \subset Z_4$, the modular ring displayed in Table 3.5.1 (previously shown in this monograph).

The class $\bar{1}_4 \subset Z_4$ has many unique features. For example, for odd integers only those in this class can equal a sum of squares and even powers as in Table 3.5.1. Because of the power restriction in regions where even powers are plentiful there is 'more room' in class $\bar{3}_4$ for primes, an anomaly noted in Riesel (1994).

All $\varphi_a(p) \in \bar{1}_4$ (p prime) could be a sum of squares. When $a = 5$, $r_1 = 1$ for a sum of squares in the well-known form [11, 12]

$$F_p = F^2_{\frac{p+1}{2}} + F^2_{\frac{p-1}{2}} \tag{3.5.3}$$

Later in this section, we shall also consider other Fibonacci sequences with this. We note that when r_1 is odd the parity structure of the sequences conforms to the pattern odd – odd – even – odd – odd – even …, but when r_1 is even we get odd only in the sequence (Table 3.5.2), a fact which follows from Table 3.5.1 and Equation (3.5.3).

When an element of any sequence is at a point where $n = p$ (prime), the class of r_1 determines the class of $\varphi_a(p)$ as in Table 3.5.3 where it is clear that the row of $a \in \bar{1}_4$ dominates the characteristics of each sequence.

Table 3.5.1. Classes and rows for Z_4

Row r_i ↓	Class i →	$\bar{0}_4$	$\bar{1}_4$	$\bar{2}_4$	$\bar{3}_4$	Comments
0		0	1	2	3	$N = 4r_i + i$
1		4	5	6	7	even $\bar{0}_4$, $\bar{2}_4$
2		8	9	10	11	$(N^n, N^{2n}) \in \bar{0}_4$
3		12	13	14	15	odd $\bar{1}_4$, $\bar{3}_4$; $N^{2n} \in \bar{1}_4$

Table 3.5.2. Parity structure of $\varphi_a(n)$ [o: odd; e: even]

a	r_1	Class of r_1	Parity structure of $\varphi_a(n)$	37	9	$\bar{1}_4$	ooe
5	1	$\bar{1}_4$	ooe	41	10	$\bar{2}_4$	ooo
13	3	$\bar{3}_4$	ooe	53	13	$\bar{1}_4$	ooe
17	4	$\bar{0}_4$	ooo	57	14	$\bar{2}_4$	ooo
29	7	$\bar{3}_4$	ooe	61	15	$\bar{3}_4$	ooe

Table 3.5.3. Class of $\varphi_a(p)$

Class of r_1	Class of $\varphi_a(p)$	a
$\bar{0}_4, \bar{1}_4$	all $\bar{1}_4$	5, 17, 37, 53, ...
$\bar{3}_4$	$\bar{1}_4\bar{3}_4\bar{1}_4\bar{3}_4$...	13, 29, 61
$\bar{2}_4$	all $\bar{3}_4$	41, 57, ...

Table 3.5.4. Factors when $p = a$

a = p	$F_{\varphi_a}(p)$	Factors
5	5	5×1
13	14209	13×1093
17	2135149	17×125597
29	77433768659591	29×2670129953779
37	3429971579923472556l	$37 \times 92701934592526285 3$

For the first member of the Golden Ratio family, $a = 5$, and when n is a prime p, $F_{\varphi_a}(p)$ can also be prime (Williams 1998). When $n \neq p$, no prime is formed except for $p = 3$. This seems to apply for all members of the Golden Ratio family of Fibonacci sequences. An interesting structural feature is that when $p = a, \varphi_a(p)$ is always composite with the smallest factor equal to a (Table 3.5.4). The prime and composite integers generated for $a = 13$ (r_1 odd) and $a = 17$ (r_1 even) are now set out in Tables 3.5.5 and 3.5.6 where

Factors of

$$N_p = kp \pm 1 \tag{3.5.4}$$

It is found that the same criteria for distinguishing primes and composites when $a = 5$ also apply for higher members of the Golden Ratio family. For $a = 5$, k is often equal to 2, and this is the case for $F_{\varphi_{13}}(19)$ in Table 3.5.5. These results suggest that (3.5.4) is general for the family. The structure of the various sequences of the Golden Ratio family suggests that only $r_1 \in \{\bar{0}_4, \bar{1}_4\}$ will have all $\varphi_a(p) \in \bar{1}_4$ (Table 3.5.3) and these can equal a sum of squares; that is, $\varphi_a(p) = x^2 + y^2$. The (x, y) couples for $\varphi_5(p)$ may be calculated from Equation (3.5.2) whereas, in general, the $\varphi_a(p)$ (x, y) couples may be calculated from (x odd, y even):

$$x, y = \tfrac{1}{2}\left(A \pm \sqrt{2F_{\varphi_a}(p) - A^2}\right), A = x + y. \tag{3.5.5}$$

The upper limit for A is $\sqrt{2F_{\varphi_a}(p)}$ and when $F_{\varphi_a}(p)$ is prime it is close to $\sqrt{2F_{\varphi_a}(p)}$.

Table 3.5.5. Factors of $N_p = kp \pm 1$ ($a = 13$)

p	$F_{\varphi_{13}}(p)$	c/p	k	Factors
3	4	c	2	2 × 2
5	19	p	–	–
7	97	p	–	–
11	2683	p	–	–
13	14209	c	84(+)	13 1093 = 84p + 1
17	399331	c	6(+1) 228(+1)	103 = 6p + 1 3877 = 228p + 1
19	2117473	c	2(–) 8(–) 20(–)	37 = 2p – 1 151 = 8p – 1 379 = 20p – 1

Table 3.5.6. Factors of $N_p = kp \pm 1$ ($a = 17$)

$F_{\varphi_{17}}(p)$	c/p	*k*	Factors
5	p	–	–
29	p	–	–
181	p	–	–
7589	p	–	–
49661	c	4(+1) 72(+1)	53 = 4*p* + 1 937 = 72*p* + 1
2135149	c	7388(+1)	17 125597 = 7388*p* + 1
14007941	c	2(–1) 19926(–1)	37 = 2*p*–1 378593 = 19926*p* – 1

Table 3.5.7. REDs for A

$\varphi_a(p)^*$	A^*
1	1, 9
3	1, 5, 9
7	3, 5, 7
9	3, 7

Moreover, A^*, the right-end-digit (RED) of A is restricted so that (x, y) estimates are easier to find (Table 3.5.7).

Composite numbers either have one (x, y) couple with common factors or the same number as the number of factors, but primes have only one (x, y) couple with no common factors (Table 3.5.8).

Table 3.5.8. Numbers of factors associated with $\varphi_{17}(p)$

p	$\varphi_{17}(p)$	Status	x, y	Factors
3	5	p	1, 2	–
5	29	p	5, 2	–
7	181	p	9, 10	–
11	7589	p	65, 58	–
13	49661	c	181, 130 85, 206	53 × 937
17	2135149	c	1165, 882 1443, 230	17 × 125597
19	14007941	c	2929, 2330 2015, 3154	37 × 378593

It is significant that the odd and even (*x*, *y*) in Table 3.5.8 are elements of the classes $\bar{1}_4$, $\bar{2}_4$, respectively, but for those $\varphi_a(p)$ with $r_1 \in \bar{3}_4$ (for example, *a* = 13, 29, 61) there are *x*, *y* couples alternatively, that is, when $\varphi_a(p) \in \bar{1}_4$. However, if the factors are in class $\bar{3}_4$, then *x*, *y* couples do not form. When there is an odd number of factors some must be in class $\bar{1}_4$ since $\bar{3}_4\bar{3}_4 \in \bar{1}_4$ but $\bar{3}_4\bar{3}_4\bar{3}_4 \in \bar{3}_4$. Thus, for $\varphi_{17}(p)$, $x = \varphi_{17}\left(\frac{1}{2}(p+1)\right)$, $y = 2\varphi_{17}\left(\frac{1}{2}(p-1)\right)$. For instance, for $\varphi_{17}(13)$, $x = \varphi_{17}(7)$, $y = 2\varphi_{17}(6)$.

Sequences structurally compatible with φ_5 have the same parity structure and r_1 of *a* is an element of class $\bar{1}_4$.

(a) The first of these is $\varphi_{37}(p)$ (Table 3.5.9). The factors are in class $\bar{1}_4$. The *x*, *y* couples are $x = \varphi_{37}\left(\frac{1}{2}(p+1)\right), y = 3\varphi_{37}\left(\frac{1}{2}(p-1)\right)$.

Table 3.5.9. *x*, *y* couples for $\varphi_{37}(p)$

p	$\varphi_{37}(p)$	Factors	Class of factors	K (3.1)	x, y
5	109(p)	–	–	–	3, 10
7	1261(c)	13 × 97	$\bar{1}_4\ \bar{1}_4$	13 = 2p – 1 97 = 14p – 1	19, 30 35, 6
11	185329(c)	241 × 769	$\bar{1}_4\ \bar{1}_4$	241 = 22p – 1 769 = 70p – 1	327, 280 423, 80
13	2295721(p)	–	–	–	1261, 840

Table 3.5.10. ***x*, *y* couples for $\varphi_{53}(p)$**

p	$\varphi_{53}(p)$	Factors	Class of factors	k (3.1)	x, y
5	209	11 × 19	$\bar{3}_4\ \bar{3}_4$	$\begin{cases} 11 = 2p+1 \\ 19 = 4p-1 \end{cases}$	–
7	3277	29 × 113	$\bar{1}_4\ \bar{1}_4$	$\begin{cases} 29 = 4p+1 \\ 113 = 16p+1 \end{cases}$	26, 51 19, 54
11	881453	331 × 2663	$\bar{3}_4\ \bar{3}_4$	$\begin{cases} 331 = 30p+1 \\ 2663 = 242p+1 \end{cases}$	–

(b) The second of these is $\varphi_{53}(p)$ (Table 3.5.10). For the smallest p values the $\varphi_{53}(p)$ integers are composite, and the factors do not always fall in class $\bar{1}_4$. Clearly, changes in the structure lead to more complex systems.

The rows of $\varphi_9, \varphi_{25} \in \bar{2}_4$ so that $\varphi_9(p), \varphi_{25}(p) \in \bar{3}_4$ and cannot form a sum of squares (Tables 3.5.8, 3.5.9). While φ_9 produces many primes, the higher a values only contain a scatter of primes (Table 3.5.11).

Thus the closer the structures of the members of the golden ratio family are to one another the more similarities there will be. For example, when $\varphi_a \in \bar{1}_4$ and $r_1 \in \{\bar{0}_4, \bar{1}_4\}$ the more similarities there will be, particularly for the primality tests developed for φ_5. The classes of $\varphi_a(n)$ and r_1 also classify the sums of squares (Tables 3.5.2, 3.5.3, 3.5.8), and the structure is paramount for analysis of the infinity of sequences arising from the equations in this section.

Table 3.5.11. Some factors of prime positioned numbers
[r_1 is a row of a]

p	$\varphi_9(p) \in \bar{3}_4$	$\varphi_{21}(p) \in \bar{1}_4$	$\varphi_{25}(p) \in \bar{3}_4$	$\varphi_{33}(p) \in \bar{1}_4$
r_1	2	5	6	8
3	3(p)	6 = 2 × 3	7(p)	9 = 3 × 3
5	11(p)	41(p)	55 = 5 × 11 5, 11 = 2p + 1	89(p)
7	43(p)	301 = 7 × 43 7, 43 = 6p + 1	463(p)	937(p)
11	683(p)	17621 = 67 × 263 67 = 6p + 1 263 = 24p – 1	35839(p)	113993 = 11 × 43 × 241 11, 43 = 4p – 1 241 = 22p – 1
13	2731(p)	136681 = 103 × 1327 103 = 8p – 1 1327 = 102p + 1	32053 = 79 × 4057 79 = 6p + 1 4057 = 312p + 1	1282969 = 261 × 491559 261 = 20p + 1 491559 = 37812p + 3
17	48691(p)	8275601(p)	25854247(p)	164643641(p)
19	174763(p)	164457461(p)	232557151 = 419 × 555029 419 = 22p + 1 555029 = 29212p + 1	1869986953(p)

Another feature of φ_5 found to be important is that the triple (F_{p+1}, F_p, F_{p-1}) has either F_{p+1} or F_{p-1} divisible by p. Furthermore, the other two members of the triple ±1 are also divisible by p, but if $p = a$, then F_p is divisible by p (Tables 3.5.4 and 3.5.12). This divisibility feature permits a check on the accuracy of elements of the sequence when approximating them by the power of the relevant golden ratio or to reduce the triples to their primitive forms (Leyendekkers and Shannon 2013, 2014).

Table 3.5.12. p'; $p'|F_p$ where +: $p'|\varphi_p(p'+1)$; –: $p'|\varphi_p(p'-1)$

p'	$\varphi_{13}(p')$	$\varphi_{17}(p')$	$\varphi_{29}(p')$	$\varphi_{37}(p')$
5	+	+	–	+
7	+	+	–	–
11	+	+	+	–
13	p	–	–	+
17	–	p	+	+
19	+	–	+	+
23	–	+	–	+
29	–	+	p	+
31	+	+	+	+

- For $a = 5, \varphi_5$, $p|F_{p+1}$ or $p|F_{p-1}$ can be predicted, but when $a > 5$ this prediction is more complex, though commonly $p|F_{p+1}$ and when a is composite the divisibility is more consistent.
- For instance, for φ_9, only $p\,|\,\varphi_9(p-1)$ occurs, while for φ_{33}, $p\,|\,\varphi_{33}(p+1)$ is preferred.
- For $\varphi_{29}(7)$ none of the triples is divisible by 7, and
- for $\varphi_{53}(13)$ none is divisible by 13.

Apparently when $r_1 = p$ (for φ_{29}, $r_1 = 7$ and for φ_{53}, $r_1 = 13$) none of the triples is divisible by the prime if it equals r_1.

The interested reader may enjoy the challenge of extending these results to examine Havil's claim that the Golden Ratio is the world's most irrational number and the first of what he calls 'awkward' numbers (Havil 2012) and whether there are similar families of meta-Fibonacci sequences such as

$$F_n = F_{n-F_{n-1}} + F_{n-1-F_{n-2}} \qquad (3.5.6)$$

with the usual initial conditions (Vajda 1989).

3.6. Fibonacci and Lucas Primes

The Fibonacci numbers are specified by the initial conditions $F_1 = F_2 = 1$ and the second order homogeneous linear recurrence relation as before

$$F_{n+1} = F_n + F_{n-1}. \qquad (3.6.1)$$

When the initial conditions are changed to 1 and 3, we get the sequence of Lucas numbers $\{L_n\}$. This equation is characterised by the ordered Fibonacci and Lucas triples $\{F_n, F_{n+1}, F_{n-1}\}$ and $\{L_n, L_{n+1}, L_{n-1}\}$ which we shall use in the analysis in this section. There are very regular patterns in the structure of the Fibonacci sequence within the modular ring Z_5 (Table 3.6.1).

Table 3.6.1. The modular ring Z_5

Class	$\bar{0}_5$	$\bar{1}_5$	$\bar{2}_5$	$\bar{3}_5$	$\bar{4}_5$
Row	$5r_0$	$5r_1+1$	$5r_2+2$	$5r_3+3$	$5r_4+4$
0	0	1	2	3	4
1	5	6	7	8	9
2	10	11	12	13	14
3	15	16	17	18	19
4	20	21	22	23	24
5	25	26	27	28	29

Note in Table 3.6.1 that the elements of the set of right-end digits $N^* = \{(0, 5), (1, 6), (2, 7), (3, 8), (4, 9)\}$ indicate the class. REDs are indicated by an asterisk. The focus in this section will be on the positional indicator, n, when n is prime, to provide useful information on F_n itself (Table 3.6.2).

Table 3.6.2. Some characteristics of F_n (* except $n = 4$)

n	Characteristics of F_n
3k	$2 \mid F_n$
4k	$3 \mid F_n$
5k	$5 \mid F_n$
6k	$8 \mid F_n$
prime	F_n can be prime
composite	F_n cannot be prime*
$p^* = 3, 7$	$p \mid F_{p+1}$
$p^* = 1, 9$	$p \mid F_{p-1}$

The triples (F_{p+1}, F_p, F_{p-1}) can be reduced to a primitive form as follows:

- When $p^* \in \{1, 9\}$, then $p \mid F_{p-1}$;
- When $p^* \in \{3, 7\}$, then $p \mid F_{p+1}$.

With $p \mid F_{p-1}$, both F_p and F_{p+1} may be reduced to $(pK \pm 1)$, $K \in Z$. Then, if we use (3.6.1) factors may be eliminated and the equation simplified (Table 3.6.4). The same applies for $p \mid F_{p+1}$ where F_p and F_{p-1} have the form $(pK \pm 1)$. For example, for $p = 43$,

$$F_p + 1 = 43(2 \times 3 \times 13 \times 307 \times 421),$$

$$F_{p+1} = 43(3 \times 89 \times 199 \times 307),$$

$$F_{p-1} - 1 = 43(3 \times 3 \times 5 \times 11 \times 41 \times 307).$$

If we eliminate the common factors 3, 43 and 307, then we get the primitive triple:

$$F'_p = 10946 = F_{21} = F_{\frac{p-1}{2}},\ F'_{p+1} = 17711 = F_{22} = F_{\frac{p+1}{2}},$$

$$F'_{p-1} = 6765 = F_{20} = F_{\frac{p-3}{2}}.$$

When a primitive represents a Lucas triple, it corresponds to a Fibonacci sum; e.g.,

$$F'_p = F_n + F_{n-2},\ F'_{p+1} = F_{n+1} + F_{n-1},\ F'_{p-1} = F_{n-1} + F_{n-3}.$$

- for $p^* = 3, 7,\ n = \dfrac{p+1}{2}$, and
- for $p^* = 1, 9,\ n = \dfrac{p+3}{2}$.

When the primitives are given by

$$(F_n, F_{n+1}, F_{n-1}),\ (L_n, L_{n+1}, L_{n-1}):$$

- for $p^* = 3, 7,\ n = \dfrac{p-1}{2}$, and
- for $p^* = 1, 9,\ n = \dfrac{p+1}{2}$.

Moreover, the last two digits of p indicate whether the primitive represents Fibonacci or Lucas triples. For instance, when $p^* = 3, 7$, (odd, odd) gives Lucas and (even, odd) gives Fibonacci, whereas when $p^* = 1, 9$, (even, odd) gives Lucas and (odd, odd) gives Fibonacci. Note also that when p_1 and p_2 are twin primes, the primitives F'_{p_1}, F'_{p_2} equal alternative Lucas / Fibonacci equivalents as in Table 3.6.3.

Table 3.6.3. Twin prime effects

Twin primes	Category	Sequences
11, 13	Prime / prime	Fibonacci / Lucas
17, 19	Prime / composite	Lucas / Fibonacci
29, 31	Prime / composite	Lucas / Fibonacci
41, 43	Composite / prime	Lucas / Fibonacci
59, 61	Prime / prime	Fibonacci / Lucas

Table 3.6.4. Class structure of primitive Fibonacci triples

Primality of F_p	p	F_p, F_{p+1}, F_{p-1}	Triple Class Structure	Primitive Fibonacci & Lucas Triples	& Class Structure in Z_5
yes	11	89 144 55	$\bar{4}_5\ \bar{4}_5$ $\bar{0}_5$	8, 13, 5 (F_6, F_7, F_5)	$\bar{3}_5\ \bar{3}_5$ $\bar{0}_5$
yes	13	233 377 144	$\bar{3}_5\ \bar{2}_5$ $\bar{4}_5$	18, 29, 11 (L_6, L_7, L_5)	$\bar{4}_5\ \bar{4}_5$ $\bar{1}_5$
yes	17	1597 2584 987	$\bar{2}_5\ \bar{4}_5$ $\bar{2}_5$	47, 76, 29 (L_8, L_9, L_7)	$\bar{2}_5\ \bar{1}_5$ $\bar{4}_5$
no	19	4181 6765 2584	$\bar{1}_5\ \bar{0}_5$ $\bar{4}_5$	55, 89, 34 (F_{10}, F_{11}, F_9)	$\bar{0}_5\ \bar{4}_5$ $\bar{4}_5$
yes	23	28657 46368 17711	$\bar{2}_5\ \bar{3}_5$ $\bar{1}_5$	89, 144, 55 (F_{11}, F_{12}, F_{10})	$\bar{4}_5\ \bar{4}_5$ $\bar{0}_5$
yes	29	514229 832040 317811	$\bar{4}_5\ \bar{0}_5$ $\bar{1}_5$	1364, 2207, 843 (L_{15}, L_{16}, L_{14})	$\bar{4}_5\ \bar{2}_5$ $\bar{3}_5$
no	31	1346269 2178309 832040	$\bar{4}_5\ \bar{4}_5$ $\bar{0}_5$	987, 1597, 610 (F_{16}, F_{17}, F_{15})	$\bar{2}_5\ \bar{2}_5$ $\bar{0}_5$

Primality of F_p	p	F_p, F_{p+1}, F_{p-1}	Triple Class Structure	Primitive Fibonacci & Lucas Triples	& Class Structure in Z_5
no	37	24157817 39088169 14930352	$\bar{2}_5\ \bar{4}_5$ $\bar{2}_5$	5778, 9349, 3571 (L_{18}, L_{19}, L_{17})	$\bar{3}_5\ \bar{4}_5$ $\bar{1}_5$
no	41	165580141 267914296 102334155	$\bar{1}_5\ \bar{1}_5$ $\bar{0}_5$	24476, 39603, 15127 (L_{21}, L_{22}, L_{20})	$\bar{1}_5\ \bar{3}_5$ $\bar{2}_5$
yes	43	433494437 701408733 267914290	$\bar{2}_5\ \bar{3}_5$ $\bar{2}_5$	10946, 17711, 6765 (F_{21}, F_{22}, F_{20})	$\bar{1}_5\ \bar{1}_5$ $\bar{0}_5$
Yes	47	2971215073 4807526976 1836311903	$\bar{3}_5\ \bar{1}_5$ $\bar{3}_5$	28657, 46368, 17711 (F_{23}, F_{24}, F_{22})	$\bar{2}_5\ \bar{3}_5$ $\bar{1}_5$
No	53	53316291173 86267571272 32951280099	$\bar{3}_5\ \bar{2}_5$ $\bar{4}_5$	271443, 439204, 167761 (L_{26}, L_{27}, L_{25})	$\bar{3}_5\ \bar{4}_5$ $\bar{1}_5$
No	59	956722026041 1548008755920 591286729879	$\bar{1}_5\ \bar{0}_5$ $\bar{4}_5$	832040, 1346269, 514229 (F_{30}, F_{31}, F_{29})	$\bar{0}_5\ \bar{4}_5$ $\bar{4}_5$
No	61	2504730781961 4052739537881 1548008755920	$\bar{1}_5\ \bar{1}_5$ $\bar{0}_5$	3010349, 2178309, 832040 (L_{31}, L_{32}, L_{30})	$\bar{4}_5\ \bar{4}_5$ $\bar{0}_5$
No	67	44945570212853 72723460248141 27777890035288	$\bar{3}_5\ \bar{1}_5$ $\bar{3}_5$	3524578, 5702887, 2178309 (F_{33}, F_{34}, F_{32})	$\bar{3}_5\ \bar{2}_5$ $\bar{4}_5$

It is well known that the Fibonacci numbers can be generalised by summing along the leading diagonals in Pascal's triangle. That is,

$$F_n = \sum_{j=0}^{\left\lfloor \frac{n+1}{2} \right\rfloor} \binom{n-j-1}{j}.$$

When $n = p$, prime this can be conveniently re-written as

$$F_p = 2 + \sum_{j=2}^{\left\lfloor \frac{pn+1}{2} \right\rfloor} \binom{p-j}{j-1}. \tag{3.6.2}$$

We shall call these numbers Pascal-Fibonacci numbers (Table 3.6.5). An equivalent sum to that in (3.6.2) is

$$\sum_{i=2}^{\left\lfloor \frac{p-1}{2} \right\rfloor} \binom{p-i}{i-1},$$

so that for each Pascal-Fibonacci number, $N_{PF}(i-1)$, along each diagonal is given by

$$N_{PF} = \binom{p-i}{i-1}.$$

For example, when $p = 17$ and $i = 4$, the third number in the sum is $N_{P_{17}}(3) = 286$. Similarly, when $p = 43$ and $i = 5$, $N_{P_{43}}(4) = 73815$. Again, when $p = 59$, the last $i = ½(p - 1) = 29$, so the 28th number in the sum is $N_{P_{59}}(28) = 30!/28!{\times}2! = 435.$

As can be seen from above in the distinction between $p^* = 3, 7$ and $p^* = 1, 9$, the class of p in Z_5 given by the REDs is critical to the structure. Therefore, we compare the Z_5 structure of the Pascal-Fibonacci numbers on this basis (Table 3.6.6). For example, for $p^* = 1$, the first numbers 9, 29, 39, 59 equal $(5r_4 + 4) \in \bar{4}_5$, the second numbers fall in class $\bar{3}_5\,(5r_3 + 3)$, and the third and fourth fall in $\bar{0}_5\,(5r_0)$.

Table 3.6.5. Pascal-Fibonacci numbers

p	F_p	
7	5,	6
11	9,	28, 35, 15
13	11,	45, 84, 70, 21
17	15,	91, 286, 495, 462, 210, 36
19	17,	120, 455, 1001, 1287, 924, 330, 45
23	21,	190, 969, 3060, 6188, 8008, 6435, 3003, 715, 66
29	27,	325, 2300, 10626, 33649, 74613, 116280, 125970, 92378, 43758, 12376, 1820, 105
31	29,	378, 2925, 14950, 53130, 134596, 245157, 319770, 293930, 184756, 75582, 18564, 2380, 120
37	35,	561, 5456, 35960, 169911, 593775, 1560780, 3108105, 4686825, 5311735, 4457400, 2704156, 1144066, 319770, 54264, 48451, 171
41	39,	703, 7770, 58905, 324632, 1344904, 4272048, 10518300, 20160075, 30045015, 34597290, 30421755, 20058300, 9657700, 3268760, 735471, 100947, 7315, 210
43	41,	780, 9139, 73815, 435897, 1947792, 6724520, 18156204, 38567100, 64512240, 84672315, 86493225, 67863915, 40116600, 17383860, 5311735, 1081575, 134596, 8855, 231
47	45,	946, 12341, 111930, 749398, 3838380, 15380937, 48903492, 124403620, 254186856, 417225900, 548354040, 573166440, 471435600, 300540195, 145422675, 51895935, 13123110, 2220075, 230230, 12650, 276
53	51,	1225, 18424, 194580, 1533939, 9366819, 45379620, 177232627, 563921995, 1471442973, 3159461968, 5586853480, 8122425444, 9669554100, 9364199760, 7307872110, 4537567650, 2203961430, 818809200, 225792840, 44352165, 5852925, 475020, 20475, 351
59	57,	1540, 26235, 316251, 2869685, 20358520, 115775100, 536878650, 2054455634, 6540715896, 17417133617, 38910617655, 73006209045, 114955808528, 151532656696, 166509721602, 166509721602, 151584480450, 113380261800, 68923264410, 33578000610, 12875774670, 3796297200, 8344518000, 131128140, 13884156, 906192, 34165, 435

The proportion of the Pascal-Fibonacci numbers in the various categories (Table 3.6.7) shows trends that are characteristic of primality. For example, the Pascal-Fibonacci composite numbers show lack of balance in the distribution of parity. The 'dominance' of the Class $\overline{0}_5$ is common to all the Pascal-Fibonacci numbers. This latter characteristic is emphasised particularly in the case of the non-prime F_p values. If various Pascal-Fibonacci numbers are added and then checked for a common factor, this could indicate primality. For instance, for $p = 19$: (17 + 120 + 455) = 592 = 37 × 16. Addition of remaining numbers in sum plus 2 yields 3589 = 37 × 97, so that F_{19} is composite and equals 37 × 113.

Table 3.6.6. Pascal-Fibonacci integers in Z_5

p	Pascal-Fibonacci Integers in Z_5								
11	$\overline{4}_5$	$\overline{3}_5$	$\overline{0}_5$	$\overline{0}_5$					
31	$\overline{4}_5$	$\overline{3}_5$	$\overline{0}_5$	$\overline{0}_5$	$\overline{0}_5$	$\overline{1}_5$	$\overline{2}_5$	$\overline{0}_5$	$\overline{0}_5$	$\overline{1}_5$
41	$\overline{4}_5$	$\overline{3}_5$	$\overline{0}_5$	$\overline{0}_5$	$\overline{2}_5$	$\overline{4}_5$	$\overline{3}_5$	$\overline{0}_5$	$\overline{0}_5$	$\overline{0}_5$
61	$\overline{4}_5$	$\overline{3}_5$	$\overline{0}_5$	$\overline{0}_5$	$\overline{1}_5$	$\overline{0}_5$	$\overline{0}_5$	$\overline{0}_5$	$\overline{0}_5$	$\overline{0}_5$
13	$\overline{1}_5$	$\overline{0}_5$	$\overline{4}_5$	$\overline{0}_5$	$\overline{1}_5$					
23	$\overline{1}_5$	$\overline{0}_5$	$\overline{4}_5$	$\overline{0}_5$	$\overline{3}_5$	$\overline{3}_5$	$\overline{0}_5$	$\overline{3}_5$	$\overline{0}_5$	$\overline{1}_5$
43	$\overline{1}_5$	$\overline{0}_5$	$\overline{4}_5$	$\overline{0}_5$	$\overline{2}_5$	$\overline{2}_5$	$\overline{0}_5$	$\overline{4}_5$	$\overline{0}_5$	$\overline{0}_5$
53	$\overline{1}_5$	$\overline{0}_5$	$\overline{4}_5$	$\overline{0}_5$	$\overline{4}_5$	$\overline{4}_5$	$\overline{0}_5$	$\overline{2}_5$	$\overline{0}_5$	$\overline{3}_5$
7	$\overline{0}_5$	$\overline{1}_5$								
17	$\overline{0}_5$	$\overline{1}_5$	$\overline{1}_5$	$\overline{0}_5$	$\overline{2}_5$	$\overline{0}_5$	$\overline{1}_5$			
37	$\overline{0}_5$	$\overline{1}_5$	$\overline{1}_5$	$\overline{0}_5$	$\overline{1}_5$	$\overline{0}_5$	$\overline{0}_5$	$\overline{0}_5$	$\overline{0}_5$	$\overline{0}_5$
47	$\overline{0}_5$	$\overline{1}_5$	$\overline{1}_5$	$\overline{0}_5$	$\overline{3}_5$	$\overline{0}_5$	$\overline{2}_5$	$\overline{2}_5$	$\overline{2}_5$	$\overline{1}_5$
19	$\overline{2}_5$	$\overline{0}_5$	$\overline{0}_5$	$\overline{1}_5$	$\overline{2}_5$	$\overline{4}_5$	$\overline{0}_5$	$\overline{0}_5$		
29	$\overline{2}_5$	$\overline{0}_5$	$\overline{1}_5$	$\overline{4}_5$	$\overline{3}_5$	$\overline{0}_5$	$\overline{0}_5$	$\overline{3}_5$	$\overline{3}_5$	$\overline{1}_5$
59	$\overline{2}_5$	$\overline{0}_5$	$\overline{0}_5$	$\overline{1}_5$	$\overline{0}_5$	$\overline{0}_5$	$\overline{0}_5$	$\overline{0}_5$	$\overline{4}_5$	$\overline{1}_5$

Table 3.6.7. Proportions of Fibonacci numbers in various classes (modulo 5)

p	even	odd	$\bar{0}_5$	$\bar{1}_5$	$\bar{2}_5$	$\bar{3}_5$	$\bar{4}_5$
7	50	50	50	50	0	0	0
11	25	75	50	0	0	0	50
13	40	60	40	40	0	0	20
17	57	43	43	43	14	0	0
19	37	63	50	12.5	25	0	12.5
23	50	50	40	20	0	30	10
29	62	38	46	15	8	23	8
31	79	21	51	14	14	7	14
37	47	53	53	41	0	0	6
41	53	47	63	5.5	10.5	10.5	10.5
43	45	55	60	15	15	0	10
47	67	33	67	19	9	5	0
53	52	48	56	8	8	8	20
59	70	30	66	14	17	10	3

When F_p is considered as a function of F_{p+1}, we have the following two cases.

(i) $p^* = 1, 9$

The Fibonacci triples for prime p are related by Simson's identity (Benjamin 2007) which may be stated in the form

$$F_p^2 = F_{p+1}F_{p-1} + 1 \tag{3.6.3}$$

and from Tables 3.6.2 and 3.6.3, they can be expressed as

$$F_p = k_1 p + 1, \quad F_{p-1} = k_2 p, \quad F_{p+1} = k_3 p + 1.$$

For $k_1, k_2, k_3 \in \boldsymbol{Z}$. From this we get

$$F_p = AF_{p+1} - 1 \tag{3.6.4}$$

in which

$$A = \frac{k_2}{k_1} = \frac{F_{p-1}}{F_p - 1}$$

A comparison of A for various values of p (Table 3.6.8a) shows that generally for primes

$$A = \frac{F_m + F_n}{F_{m+2} - F_n}$$

where $m = ½(p + 1)$ and $n = m - 2$, whereas non-primes have a pattern

$$A = \frac{F_m}{F_n}$$

where $m = ½(p - 1)$ and $n = ½(p + 1)$.

(ii) $p^* = 3, 7$

From Tables 3.6.2 and 3.6.3, we can also obtain

$$F_p = k_1 p - 1, \quad F_{p-1} = k_2 p + 1, \quad F_{p+1} = k_3 p.$$

For $k_1, k_2, k_3 \in \boldsymbol{Z}$.

From this we get

$$F_p = AF_{p+1} + 1 \tag{3.6.5}$$

in which

$$A = \frac{k_2 p + 1}{k_1 p} = \frac{F_{p-1}}{F_p + 1}$$

As in (i), primes have

$$A = \frac{F_m + F_n}{F_{m+2} - F_n}$$

where $m = ½(p + 1)$ and $n = m – 2$, and primes also have a pattern

$$A = \frac{F_m}{F_{m+1}}$$

where $m = ½(p – 1)$. A comparison of A for various values of p is displayed below (Table 3.6.8b).

Table 3.6.8(a). *A* for *p =1, 9**

p	A as a Numerical fraction	 Fibonacci or Lucas ratio	F_{p+1}	F_p
11	5/8	F_5/F_6	144	89
19	2×17/5×11	F_9/F_{10}	6765	4181
29	3×281/4×11×31	L_{14}/L_{15}	832040	514229
31	2×5×61/3×7×47	F_{15}/F_{16}	2178309	1346269
41	7×2161/4×29×211	L_{20}/L_{21}	267914296	165580141
59	514229/5×8×11×31×61	F_{29}/F_{30}	1548008755920	956722026041
61	2×3×3×41×2521/3010349	L_{30}/L_{31}	4052739537881	2504730781961

Table 3.6.8(b). ***A* for $p^* = 3, 7$**

p	A as a numerical fraction	Fibonacci or Lucas ratio	F_{p+1}	F_p
3	1/3	L_1/L_2	3	2
7	4/7	L_3/L_4	21	13
13	8/13	F_6/F_7	377	233
17	21/34	F_8/F_9	2584	1597
23	199/14×23	L_{11}/L_{12}	46368	28657
37	8×17×19/113×37	F_{18}/F_{19}	39088169	24157817
43	4×29×211/3×43×307	L_{21}/L_{22}	701408733	433494437
47	461×139/2×47×1103	L_{23}/L_{24}	4807526976	2971215073
53	521×233/2×53×1853	F_{26}/F_{27}	86267571272	53316291173

In contrast to the primitive triples for $p^* = 3, 7$, the last two digits of p are (even, odd) for Lucas and (odd, odd) for Fibonacci. For $p^* = 1, 9$, the last two digits linked to Lucas and Fibonacci are the same as the primitive triple one.

There are also two related cases of Fibonacci squares.

(a) F_{2p+1}

When we combine Simson's identity with the equally well-known (Livio 2002)

$$F_n^2 + F_{n+1}^2 = F_{2n+1} \tag{3.6.6}$$

we get for $n = p$ (odd):

$$F_{p+1}F_{p-1} + F_{p+1}^2 = F_{2p+1} - 1 \tag{3.6.7}$$

When $p^* \in \{3, 7\}$, $p \mid F_{p+1}$ (Table 3.6.2), and so from (3.6.7) $p|(F_{2p+1} - 1)$ (Table 3.6.9a).

Obviously, F_{2p+1} may be obtained directly from (3.6.9) so that when $2p+1$ is prime this gives additional information on the production of primes. Note that for this set $F_{2p+1}-1 = pQ$, where $Q \in \bar{4}_5$ for $p^* = 3$, and $Q \in \bar{2}_5$ for $p^* = 7$. When $p^* \in \{1, 9\}$, $p \mid F_{p-1}$ (Table 3.6.2), and so from (3.6.6) and (3.6.7) $p \mid (F_{2p-3} - 1)$ (Table 3.6.9b). Note that for this set $(F_{2p-3} - 1) = pQ$, where Q is even and $\in \bar{0}_5$ for $p^* = 1$, and $Q \in \bar{1}_5$ for $p^* = 9$.

(b) Class patterns of p^2 and F_p^2

Since $F_p^2{}^* = p^2{}^* F_p^2$ and p^2 are in the same class, then $\left(F_p^2 - p^2\right) \llcorner \bar{0}_5$ (Table 3.6.10). For $p^* \in \{3, 7\}$, $F_p^2 - p^2$ has 120 as the last three digits (Classes $\bar{2}_5$, $\bar{3}_5$), whereas for $p^* \llcorner \{1,9\}$, $F_p^2 - p^2$ has X00 as the last three digits (Classes $\bar{1}_5$, $\bar{4}_5$).

Table 3.6.9(a). $p \mid (F_{2p+1} - 1)$

p	2p + 1	$F_{2p+1} - 1$		
13	27	196417	=13×15109	=13(29×521)
17	35	9227464	=17×542792	=17(2³x19x3571)
23	47	2971215072	=23×129183264	= 23(2⁵x3³×7×139×461)
37	75	2111485077978649	= 37×57067164269677	= 37(57067164269677)

Table 3.6.9(b). $p \mid (F_{2p-3} - 1)$

p	2p – 3	$F_{2p-3} - 1$		
11	19	4180	= 11×380	= 11(2²×5×19))
19	35	9227464	= 19×485656	= 19(2³×17×3571)
29	55	139583862444	= 29×4813236636	= 29(2²×3×13×19×281×5779)
31	59	956722026040	= 31×30862000840	= 31(2³×5×8×11×59×19489)

Table 3.6.10. $F_p^2 - p^2$

p	F_p	F_p^2	p^2	$F_p^2 - p^2$
5	5	25	25	0
7	13	169	49	120
11	89	7921	121	7800
13	233	54289	169	54120
17	1597	2550409	289	2550120
19	4181	17480761	361	17480400
23	28657	821223649	529	821223120
29	514229	264431464441	841	264431463600
31	1346269	1812440220361	961	1812440219400
37	24157817	582413122205489	1369	582413122204120

For composites only $X = 4$ in the range considered. This difference in structure might be another indication of primality. (For $p = 41, 43$, $F_p^2 - p^2$ ends in 8200 and 5120, respectively.)

Since p is odd, the influence of F_{p+2}, F_{p+3} is further seen by

$$F_{p-2}F_{p+2} = F_p^2 + 1 \tag{3.6.8}$$

and

$$F_{p-3}F_{p+3} = F_p^2 - 4 \tag{3.6.9}$$

so that from Simson's identity and (3.6.8), we get

$$F_{p-1}F_{p+1} = F_{p-2}F_{p+2} - 2. \tag{3.6.10}$$

Thus $p \mid \left(F_{p-2}F_{p+2} - 2\right)$ (Table 3.6.11), in which 24 is always a factor of $\left(F_{p-2}F_{p+2} - 2\right)$.

Table 3.6.11. $((F_{p-2} \times F_{p+2}) - 2)/p$

p	F_p	F_{p-2}	F_{p+2}	$((F_{p-2} \times F_{p+2}) - 2)/p$
7	13	5	34	24
11	89	34	233	24(2×15)
13	233	89	610	24(2×87)
17	1597	610	4181	24(7×893)
19	4181	1597	10946	24(5×11×697)
23	28657	10946	75025	24(2^2×3×7×89×199)
29	514229	196418	1346269	24(5×11×13×31×61×281)
31	1346269	514229	3524578	24(5×19×331×77471)
37	24157817	9227465	39088169	24(2×3×17×19×339116277)

The last remark is a consequence of the opposite parity of F_{p-1} and F_{p+1} with 3 as a factor of either and 8 as a factor of the even number (Table 3.6.12).

Table 3.6.12. 'X' divisible by *p*

p	F_{p+1}	Factors of		F_{p-1}
		F_{p+1}	F_{p-1}	
7	21	3, X	8	8
11	144	3, 8	X	55
13	377	X	3, 8	144
17	2584	8, X	3	987
19	6765	3	8, X	2584
23	46368	3, 8, X	–	17711
29	832040	8	3, X	317811
31	2178309	3	8, X	832040
37	39088169	X	3, 8	14930352
41	267914296	8	3, X	102334155
43	701408733	3, X	8	267914296
47	4807526976	X	3, 8	1836311903
53	86267571272	8, X	3	32951280099
59	1548008755920	3, 8	X	591286729879
61	4052739537881	–	3, 8, X	1548008755920

The distribution of 3 and 8 suggests that some composites have F_{p-1} even and divisible by 8 as well as p. Many cases have the factors 3 and 8 confined to one of the numbers adjacent to F_p. However, p^* effects are overlaid. Comparison of the factors of F_{p-2} and F_{p-1} shows that none are common. In fact, for the composites F_{19} and F_{31}, F_{p-2} values are prime. However, the prime F_{13} also has this feature. Comparison of F_{p-2} and F_{p-3} yields the same result.

The variety of structural patterns presented here show that differences exist between Fibonacci prime-suffixed index numbers, F_p, which produce prime numbers, and those which produce composite numbers. While the overall structure of F_p is stable, as can be inferred from the tables in this section, there is no reason, at least from a structural point of view, why there should not be an infinity of Fibonacci primes. Further analysis using different modular rings should be useful including the more general Z_p, considered as the ring of p-adic generalized integers (Araci et al. 2013; Kim et al. 2013). These results can be used for analyzing the generalized Fibonacci numbers associated with the golden ratio family in Section 3.5.

3.7. Prime Sequences from an Extended Sophie Germain Model

When p and $2p+1$ are both prime, p is a Sophie Germain prime (Ireland and Rosen 1990). Sophie Germain proved in the 1820s that the first case of ***Fermat's last theorem*** is true for such primes (Edwards 1977). It is not known if there is an infinity of such primes (Hoffman 1998). Sequences $\{N_n\}$ can be generated by continuing this algorithm:

$$N_n = 2^{n-1} p + \left(2^{n-1} - 1\right) \tag{3.7.1}$$

with p, not always a Germain prime, the initial value, or equivalently for calculating:

$$N_n = 2^{n-1}(p+1)-1 \tag{3.7.2}$$

with the former slightly more interesting because of the obvious presence of a Mersenne prime.

A limitation of the sequences is that only primes in class $\bar{3}_4$ of the modular ring Z_4 will be produced (Leyendekkers et al 2007) (Table 3.7.1).

Another limitation is that starting primes that have appeared in a previous sequence will only produce repeat sequences, somewhat like, but different from, the case of the Collatz sequences. Thus, ½(p – 1) is not suitable for new sequences if it is a prime since this prime will have yielded a sequence (Table 3.7.3 and 3.7.4).

(a) Class of N_n

If $p = (4r_1+1) \in \bar{1}_4$, then

$$2(4r_1+1)+1 = 4(2r_1)+3, \tag{3.7.3}$$

or if $p = (4r_3+3) \in \bar{3}_4$, then

$$2(4r_3+3)+1 = 4(2r_3+1)+3, \tag{3.7.4}$$

so that $\{N_n\} \in \bar{3}_4$ or, more simply, $a_n p \in \bar{0}_4$ and $b_n \in \bar{3}_4$; thus $(a_n p + b_n) \in \bar{3}_4$, where

$$a^n = 2^{n-1},$$

$$\begin{aligned} b_n &= 2^{n-1} - 1 \\ &= \bar{0}_4 - 1 \\ &= 4(r_0 - 1) + 3. \end{aligned}$$

(b) Right-end-digits (REDs)

An N_n with a RED, (N_n^*), equal to 5 (apart from 5 itself) cannot be a prime. This RED occurs for all n given by $n = s + 4t$ except when N_n^*=9 (Table 2). With p^* = 9 we get 2x9 + 1 = 19, so that $N_n^* \neq 5$.

Table 3.7.1. Classes and rows for Z4

Row $r_i \downarrow$	Class $i \rightarrow$	$\bar{0}_4$	$\bar{1}_4$	$\bar{2}_4$	$\bar{3}_4$	Comments
0		0	1	2	3	$N = 4r_i + i$
1		4	5	6	7	even $\bar{0}_4, \bar{2}_4$
2		8	9	10	11	$(N^n, N^{2n}) \in \bar{0}_4$
3		12	13	14	15	odd $\bar{1}_4, \bar{3}_4; N^{2n} \in \bar{1}_4$

Table 3.7.2. N_n^*=5 when $n = s + 4t$

$p^* \downarrow ; n \rightarrow$	1	2	3	4	5	s
1	1	3	7	5	1	0
3	3	7	5	1	3	3
7	7	5	1	3	7	2
9	9	9	9	9	9	---

(c) Rows of N_n

From equation (3.7.3) equation (3.7.2) is effectively

$$N_n = 4r_3 + 3$$

so that

$$\begin{aligned} r_3 &= \tfrac{1}{4}\left(2^{n-1}(p+1)\right) - 1 \\ &= 2^{n-3}(p+1) - 1 \qquad (3.7.5) \\ &= N_{n-2}. \end{aligned}$$

Thus for a prime each prime will be repeated in the rows. For instance, when $p = 19$, $N_n = 1279$, and $N_9 = 4\text{x}1279 + 3 = 5119$ with $r_3 = 1279$.

(d) Digit Structure of N_n

For $N_n^* = 9$, the integers have the structure

$$N_n = |A|9 \text{ or} |A|99 \text{ and } N_{n+1} = |2A+1|9 \text{ or} |2A+1|99, \ldots$$

and the other REDs of p have 1,3,5,7 replacing 9.

Since $N_n^* = 9$, $N_n^* \neq 5$ in the series, and it could be expected that more primes would be produced (Tables 3.7.3(a) and (b)).

Some examples for $N_n^* = 1,3,7$ are shown in Tables 3.7.4(a) and (b).

For a given n the coefficient b_n is the same for all primes. When $(n - 1) = p$, b_n can be a Mersenne prime as we saw in equation (3.7.1). Some examples are set out in Table 3.7.5 in which the asterisked primes are from the sequence with $p = 31$ in Table 3.7.4.

Table 3.7.3(a). $N^* = 9; 1 \leq n \leq 8$

p↓;n→	1	2	3	4	5	6	7	8
19	19	---	79	---	---	---	1279	---
29	29	59	---	239	479	---	---	---
59	29	59	---	239	479	---	---	---
89	89	179	359	719	1439	2879	---	11519
109	109	---	439	---	1759	---	7039	---
139	139	---	---	---	2239	---	---	---
149	149	---	599	---	2399	4799	---	---
179	89	179	359	719	1439	2879	---	11519
199	199	---	---	---	---	---	12799	---
229	229	---	919	---	---	---	---	---
239	239	479	---	---	---	---	15359	---
269	269	---	---	---	---	---	---	---
349	349	---	1399	---	---	---	---	---
359	89	179	359	719	1439	2879	---	11519
379	379	---	---	---	6079	---	---	---
389	389	---	1559	3119	---	12479	---	49919

Table 3.7.3(b). $N^* = 9; 9 \leq n \leq 16$

p↓;n→	9	10	11	12	13	14	15	16
19	5119	---	20479	---	81919	---	---	---
29	---	15359	---	---	---	245759	---	---
59	---	15359	---	---	---	245759	---	---
89	23039	---	---	---	---	---	1474559	2949119
109	---	---	---	---	---	---	1802239	---
139	35839	---	---	---	---	---	---	---
149	---	---	---	---	---	---	---	---
179	23039	---	---	---	---	---	1474559	2949119
199	5119	---	---	---	---	---	3276799	---
229	---	---	235519	---	942079	1884159	---	---
239	---	---	---	492319	---	---	---	7877119
269	69119	138239	---	---	1105919	---	---	8847359
349	89599	---	---	---	---	---	---	---
359	23039	---	---	---	---	---	1474559	2949119
379	---	---	---	---	---	---	---	---
389	99839	199679	---	---	---	3194879	---	---

Table 3.7.4(a). $N^* \varepsilon \{1,3,7\}$; $1 \le n \le 8$

$p\downarrow;n\rightarrow$	1	2	3	4	5	6	7	8
5	5	11	23	47	---	191	383	---
7	7	123	---	---	---	223	---	---
11	5	11	23	47	---	191	383	---
13	7	123	---	---	---	223	---	---
17	17	---	71	---	---	---	1151	---
19	19	---	79	---	---	---	1279	---
23	5	11	23	47	---	191	383	---
29	29	59	---	239	479	---	---	---
31	31	---	127	---	---	---	---	---
137	137	---	---	1103	2207	---	8831	---

Table 3.7.4(b). $N^* \varepsilon \{1,3,7\}$; $9 \le n \le 16$

$p\downarrow;n\rightarrow$	9	10	11	12	13	14	15	16
5	---	---	6143	---	---	---	---	---
7	---	3583	---	---	---	---	---	---
11	---	---	6143	---	---	---	---	---
13	---	3583	---	---	---	---		---
17	---	---	---	---	73727	---	294911	---
19	5119	---	20479	---	81919	---	---	---
23	---	---	6143	---	---	---	---	---
29	---	15359	---	---	---	245759	---	---
31	8191	---	---	---	131071	---	524287	---
137	35327	---	141311	---	56527	---	---	---

Table 3.7.5. $b_{n'} = M_p$

n'	3	4	6	8	14	18	20
$n' - 1$	2	3	5	7	13	17	19
$2^{n'-1} - 1$	3	7	31*	127*	8191*	131071*	524287*

When $p = 31 = (2^5 - 1)$, (3.7.2) becomes

$$N_n = 2^{n+4} - 1$$
$$= 2^{n'-1} - 1$$

when $n = n' - 5$. This has the form of a Mersenne number, and this occurs whenever $(p+1) = 2^m$.

By way of conclusion to this section, we note that the primes 89 (Table 3.7.3) and 137 (Table 3.7.4), as well as giving good prime yields, are distinctive in the following ways. 89 is a Fibonacci prime, and

$$\frac{1}{89} = \sum_{k=1}^{\infty} F_k \times 10^{-k-1} = 0.011235\ldots\text{(Livio 2002).} \tag{3.7.6}$$

1/137 is approximately the fine-structure constant (Sommerfeld's constant), α, a fundamental physical constant which characterizes the strength of the electromagnetic interaction between elementary charged particles (Stakhon et al 2017). Leahy (1996) relates α to the Fibonacci numbers through the sequence generated by

$$L = \sum_{k=1}^{\infty} (6k - 0.5). \tag{3.7.7}$$

At what level these can seem like idle curiosities, but these coincidences have puzzled generations of physicists [6].

Finally with (3.7.1) in the form

$$N_n = a_n p + b_n \tag{3.7.8}$$

a_n is even and can be written as a sum of primes (Livio 2002), so that when b_n is a Mersenne prime and N_n, is prime (Table 3.7.6).

Table 3.7.6. $2^{n-1} \times 89$

n	$N_n = p$	a_n89	$b_n = M_p$	Σprimes = p
3	359	356 = 353 + 3	3	353 + 3 + 3
4	719	712 = 701 + 11	7	701 + 11 + 7
8	11519	11392 = 11369+23	127	11369+127+23

Chapter 4

FIBONACCI NUMBERS AND THE GOLDEN RATIO FAMILY

The Golden Ratio can be considered as the first member of a family which can generate a set of generalized Fibonacci sequences. Here we consider some related problems in terms of the Binet form of these sequences, $\{F_n(a)\}$, where the sequence of ordinary Fibonacci numbers can be expressed as $\{F_n(5)\}$ in this notation. A generalized Binet equation can predict all the elements of the Golden Ratio family of sequences. Identities analogous to those of the ordinary Fibonacci sequence are developed as extensions of work by Filipponi, Monzingo and Whitford in *The Fibonacci Quarterly*, by Horadam and Subba Rao in the *Bulletin of the Calcutta Mathematical Society*, within the framework of Sloane's *Online Encyclopedia of Integer Sequences*.

It is proved in the first section of this chapter that infinite sequences of generalized Fibonacci sequences obtained from generalizations of the Golden Ratio can generate all primitive Pythagorean triples. This is a consequence of the integer structure since the major component of a primitive Pythagorean triple always has the form $(4r_1 + 1)$ where r_1 belongs to the class in the modular ring Z_4. The decimal expansion of the Golden Ratio is then examined through the use of various properties of the Fibonacci numbers and some exponential functions. Research into properties of generalizations of the Golden Ratio has also considered various forms of extreme and mean ratios. This is considered here too within the framework of a family of surds, $\frac{1}{2}\sqrt{1+a}$, and generalized Fibonacci numbers, $F_n(a)$, with the ordinary Fibonacci numbers being the particular case when $a = 5$. We refer to this connection in the last section of this chapter which deals with the Collatz Conjecture, or the $3n+1$ conjecture, which has been around for about eighty years.

4.1. Primitive Pythagorean Triples and Generalized Fibonacci Sequences[5]

In the notation of the Fibonacci sequences of the Golden Ratio family, the generalized Fibonacci numbers satisfy the second order recurrence relation

$$F_{n+1}(a) = F_n(a) + r_1 F_{n-1}(a) \qquad (4.1.1)$$

with unity as the two initial terms, and in which r_1 is the row of the variable a for $\phi(a)$ of the Golden Ratio family, with

$$\varphi_a = \frac{1+\sqrt{a}}{2} \qquad (4.1.2)$$

and (Table 4.1.1).

[5] Based on Leyendekkers and Shannon (2017).

Table 4.1.1. Classes and rows for Z_4

Row $r_i \downarrow$	Class $i \rightarrow$	$\bar{0}_4$	$\bar{1}_4$	$\bar{2}_4$	$\bar{3}_4$	Comments
0		0	1	2	3	$N = 4r_i + i$
1		4	5	6	7	even $\bar{0}_4, \bar{2}_4$
2		8	9	10	11	$\left(N^n, N^{2n}\right) \in \bar{0}_4$
3		12	13	14	15	odd $\bar{1}_4, \bar{3}_4$; $N^{2n} \in \bar{1}_4$

The generalized Binet formula in this notation is then (Shannon and Leyendekkers, 2015)

$$F_n(a) = \frac{\left(\dfrac{1+\sqrt{a}}{2}\right)^n - \left(\dfrac{1-\sqrt{a}}{2}\right)^n}{\sqrt{a}}. \tag{4.1.3}$$

which is well-known for the ordinary Fibonacci numbers as

$$\varphi_a = \frac{1+\sqrt{a}}{2} \tag{4.1.4}$$

and any Golden Ratio family member is given by

$$\frac{F_n(a)}{F_{n-1}(a)} \rightarrow \varphi_a.$$

Other approaches to what has been called "generalized Golden numbers" are in the literature (Hongquan et al. 1996; Moore 1993), but they are more like generalized Pell numbers, which are in effect covered in (Shannon and Horadam 1994). The purpose of this section is then to demonstrate that all Pythagorean triples can be formed from these Golden Ratio generalized Fibonacci numbers

For the ordinary Fibonacci numbers

$$F_{2n+1}(5) = F_{n+1}^2(5) + F_n^2(5). \tag{4.1.5}$$

More generally this becomes

$$F_{2n+1}(a) = F_{n+1}(a)^2 + r_1 F_n(a)^2. \tag{4.1.6}$$

For example, when $r_1 = 1$ and $n = 5$ (4.1.6) becomes (4.1.5):

$$\begin{aligned} F_{11}(5) &= 89 \\ &= 64 + 25 \\ &= F_6(5)^2 + 1 \times F_5(5)^2. \end{aligned},$$

Table 4.1.2 displays some of the first few terms of some of these $\{F_n(a)\}$.

Equation (4.1.6) can be extended with (4.1.1) to

$$F_n(a)^2 + F_{n+1}(a)^2 = 2F_n(a)F_{n+1}(a) + r_1^2 F_{n-1}(a)^2. \tag{4.1.7}$$

For example, when $r_1 = 1$ and $n = 5$ (4.1.7) becomes

$$\begin{aligned} F_5(5)^2 + F_6(5)^2 &= 25 + 64 \\ &= 89 \\ &= 80 + 9 \\ &= 2F_5(5)F_6(5) + 1^2 F_4(5)^2. \end{aligned}$$

Table 4.1.2. $F_n(a)$, $n = 1, 2, \ldots, 12$

r_1	n	*1*	*2*	*3*	*4*	*5*	*6*	*7*	*8*	*9*	*10*	*11*	*12*
0	$F_n(1)$	1	1	1	1	1	1	1	1	1	1	1	1
1	$F_n(5)$	1	1	2	3	5	8	13	21	34	55	89	144
2	$F_n(9)$	1	1	3	5	11	21	43	85	171	341	683	1365
3	$F_n(13)$	1	1	4	7	19	40	97	217	508	1159	2683	6160
4	$F_n(17)$	1	1	5	9	29	65	181	441	1165	2929	7589	19305
5	$F_n(21)$	1	1	6	11	41	96	301	781	2286	6191	17621	48576
6	$F_n(25)$	1	1	7	13	55	133	463	1261	4039	11605	35839	105469
7	$F_n(29)$	1	1	8	15	71	176	673	1905	6616	19951	66263	205920
8	$F_n(33)$	1	1	9	17	89	225	937	2737	10233	32129	113993	371025

Table 4.1.3. Examples of $\{N_i\}$ sequences and associated primitive Pythagorean triples

a	r_1	$\{N_i\}$	Primitive Pythagorean triples from odd N_i
13	3	17,65,410,1961, 11009,...	(17,15,8);(65,56,33);(1961,1520,1239); (11009,7809,7760)
21	5	37,157,1802,10897, 99817,700562,...	(37,35,12);(157,132,85);(10897,77535,787); (99817,81385,57792)
29	7	65,289,5266,36017, 483905,4081954,...	(65,63,16);(289,240,161);(36017,25935,24); (483905,143165,236896)
45	11	145,673,24554,190489, 44631233,40476770,...	(145,143,24); (673,552,385); (190489,126480,42439)

Equation (4.1.6) only gives numbers already in the sequence, but for $a > 5$

$$N_n = F_{n+1}(a)^2 + F_n(a)^2. \tag{4.1.8}$$

will yield new sequences for each value of a. When $F_n(a), F_{n+1}(a)$ have opposite parity the sequences have an odd, even, odd pattern. In this case N can be the major component of a Pythagorean triad (Krishna 1974) in which the minor components will be $2F_{n+1}(a)F_n(a)$ and $\left(F_{n+1}(a)^2 - F_n(a)^2\right)$ (Table 4.1.3, in which the $\{N_i\}$ values are calculated from (4.1.8) and Table 4.1.2).

When r_1 is even all the $F_n(a)$ are odd so that the sum of squares will be even. However, division by 2 can give odd values and these can be used to form primitive Pythagorean triples (pPts). For example, when $r_1 = 2$, the sequence is

$$\{M_n\} \equiv \tfrac{1}{2}\{N_n\} \equiv \{5,17,73,281,...\} \subset \bar{1}_4.$$

The elements of $\{M_n\}$ can all be the major component of a pPt:

$c = x^2 + y^2.$

The values of *x* and *y* can be estimated from

$$x, y = \frac{A \pm \sqrt{2M_n - A^2}}{2} \tag{4.1.9}$$

in which *x* is odd and *y* is even with $A = x + y$. Then $a = x^2 - y^2, b = 2xy$. But

$$2M_n = F_{n+1}(a)^2 + F_n(a)^2 \text{ and } A = F_{n+1}(a)$$

so that

$$x, y = \tfrac{1}{2}\left(F_{n+1}(a) \pm F_n(a)\right).$$

For instance, when $r_1 = 8$,

$F_6(33) = 225, F_7(33) = 937, M_6 = 464297,$ and

$x = \tfrac{1}{2}(937 + 225) = 581,$ and

$y = \tfrac{1}{2}(937 - 225) = 356,$ so that

$c = 581^2 + 356^2 = 464297,$

$b = 581^2 - 356^2 = 210825,$

$a = 2(581 \times 356) = 413672,$

and

$c^2 = 215,571,704,209,$

$b^2 = 44,447,180,625$

$$a^2 = 171,124,523,584.$$

Thus it can be readily confirmed that the components may also be calculated directly from the generalized Fibonacci numbers by

$$c = \tfrac{1}{2}\left(F_{n+1}(a)^2 + F_n(a)^2\right) \qquad (4.1.10)$$

$$b = F_{n+1}(a)F_n(a) \qquad (4.1.11)$$

$$a = \tfrac{1}{2}\left(F_{n+1}(a)^2 - F_n(a)^2\right), \qquad (4.1.12)$$

so that $c^2 = a^2 + b^2$ becomes

$$\left(\tfrac{1}{2}\left(F_{n+1}(a)^2 + F_n(a)^2\right)\right)^2 = \left(\tfrac{1}{2}\left(F_{n+1}(a)^2 - F_n(a)^2\right)\right)^2 + \left(F_{n+1}(a)F_n(a)\right)^2,$$

which accords with known results (Horadam 1961; Shannon and Horadam 1973) for generalized $F_n(5)$ and which appear in another form later in this section. Some examples appear in Table 4.1.4.

Table 4.1.4. Examples of $\{M_i\}$ sequences and associated primitive Pythagorean triples

a	r_1	$\{M_i\}$	Primitive Pythagorean triples from M_i
9	2	5,17,73,281,1145,4537, 18233, ...	(5,4,3);(17,15,8);(73,554,48);(281,231,160); (1145,1064,423); (4537,3655,26880);(18233,14535,11008).
33	8	41,185,4105,29273, 464297,4184569, 56102729, ...	(41,40,9);(185,176,57);(4105,3816,1513); (29273,21352,20025); (464297,413672,210825);(4184569,3306600,2564569); (56102729,48611560,28007721).

This establishes the result that all Pythagorean triples are generalized Fibonacci triples generated from the generalized Golden Ratio. We now expand this theme.

We have shown that the major component of all primitive Pythagorean triples (pPts) ($c^2 = a^2 + b^2$) must fall in Class $\bar{1}_4 \in Z_4$, a modular ring, (Table 4.1.1); that is

$$c = 4r_1 + 1 \tag{4.1.13}$$

since only odd numbers in this class are sums of squares: $c = x^2 + y^2$. The generalized Golden Ratio family members are given by

$$\varphi = \tfrac{1}{2}\left(1 + \sqrt{a}\right) \tag{4.1.14}$$

where

$$a = 4r_1 + 1$$

and hence

$$a = (2\varphi - 1)^2. \tag{4.1.15}$$

This means that the major components of all primitive Pythagorean tripless may be obtained from the generalized Golden Ratio φ_a.

Thus, we have not only displayed the role of the generalized Golden Ratio and thus extended the results of Horadam, but also provided insight into the integer structure through the associated modular ring.

When a is prime, all $a = x^2 + y^2$ are the major components of primitive Pythagorean triples. In this case x and y have no common factors.

Moreover, composites that only have factors in Class $\bar{3}_4$ can never be equal to a sum of squares. All other composites in $\bar{1}_4$ equal either one (*x*, *y*) couple with a common factor or have the same number of (*x*, *y*) couples as factors (Table 4.1.5).

Table 4.1.5. Factors, couples and pPts

a	Factors	Class	x,y	$(2\varphi-1)^2$ pPts
9	3,3	$\bar{3}_4\,\bar{3}_4$	---	---
17	prime	$\bar{1}_4$	1,4	17,15,8
25	5,5	$\bar{1}_4\,\bar{1}_4$	3,4	25,7,24**
41	prime	$\bar{1}_4$	5,4	41,40,9
65	5,13	$\bar{1}_4\,\bar{1}_4$	1,8	65,63,16
			7,4	65,56,33
101	prime	$\bar{1}_4$	1,10	101,99,20
237	3,79	$\bar{3}_4\,\bar{3}_4$	---	---
333	3,3,37	$\bar{3}_4\,\bar{3}_4\,\bar{1}_4$	3,18	333,315,108***
577	prime	$\bar{1}_4$	1,24	577,575,48
733	prime	$\bar{1}_4$	2,27	733,725,108
3725	5,5,149	$\bar{1}_4\,\bar{1}_4\,\bar{1}_4$	35,50	3725,1275,3500
13297	prime	$\bar{1}_4$	79,84	13297,815,13272
69589	13,53,101	$\bar{1}_4\,\bar{1}_4\,\bar{1}_4$	183,190	69589,2611,69540
			217,150	69589,24589,65100
			105,242	69589,47539,50820
1401953	7,19,83,127	$\bar{3}_4\,\bar{3}_4\,\bar{3}_4\,\bar{3}_4$	---	---

[**Squares have only one (x, y); *** x and y have a common factor]

The (x, y) couples may be calculated from

$$x, y = \frac{A \pm \sqrt{2a - A^2}}{2} \tag{4.1.16}$$

in which x is odd and y is even with $A = x + y$. For primes $A \sim \sqrt{2a}$ and this is the maximum value otherwise.

From Equation (4.1.14) we can construct Table 4.1.6 for $n = 14$:

Table 4.1.6. $c \sim a = \left((2F_{n+1} - F_n)/F_n\right)^2$

r_1	1	2	3	4	5	6
c	5.0	9.0	13.0	17.0	20.7	22.1
a	5	9	13	17	21	25

so that we need $n > 14$; for instance, as in Table 4.1.7.

Table 4.1.7. Approximation of F_{20}/F_{19} to $\frac{1+\sqrt{a}}{2}$

r_1	a = c	F_{20}	F_{20}/F_{19}	$\frac{1+\sqrt{a}}{2}$
1	5	6765	1.6180339	1.6180339
2	19	349525	1.9999942	2.0000000
3	13	4875913	2.3027037	2.3037756
4	17	35877321	2.5612022	2.5615528
5	21	179854741	2.7902858	2.7912878

In summary, $a = c$ for all primes since they equal $x^2 + y^2$ and for all composites provided they have factors in class $\bar{1}_4$. In this case there will be the same number of the x, y pairs as factors, so multiple pPts can be obtained. For instance, $a = c = 65$ and $65 = 1^2 + 8^2 = 4^2 + 7^2$, and there are

2 pPts. Note that the first and third rows of Table 4.1.3 yield (65, 56, 33) and (65, 63, 16); and (65, 56, 33) = (65, (2 × 28), (49 – 16)) and (65, (64 – 1), (2 × 8)). The Table 4.1.3 values can be obtained from a = 65 directly (Table 4.1.5) so there is duplication in the process.

Pythagorean triples have been studied since antiquity but this section shows that we only need a to get all pPts. The results here generalize alternative methods (Horadam 1966; Horadam and Shannon 1973) which in effect only duplicate those from a and hence $\bar{1}_4 \in Z_4$. More generally though, since $a \in \bar{1}_4$ it can equal $x^2 + y^2$ which is essential for c.

When a is prime there is always an (x, y) pair, but if a is composite there must be factors in class $\bar{1}_4$, otherwise there is no (x, y) pair. For example,

$$65 = 5 \times 13,\ 5{,}13 \in \bar{1}_4,$$
$$65 = 1^2 + 8^2 = 4^2 + 7^2,$$

and there is the same number of pairs of squares as there are factors. Thus c must equal

$$a = x_1^2 + y_1^2 = 4r_1 + 1 \in \bar{1}_4 \subset Z_4.$$

It follows that all pPts have components obtained from generalized Fibonacci number triples. Since

$$\varphi = \tfrac{1}{2}\left(1 + \sqrt{a}\right),$$
$$a = (2\varphi - 1)^2.$$

Thus, for sufficiently large n, and provided $a = x^2 + y^2$,

$$c = \left(2\varphi - 1^2\right) = \left(\left(2F_{n+1}(a) - F_n(a)\right)/\left(F_n(a)\right)\right)^2$$

Table 4.1.8. $L_n(a)$, $n = 1, 2, \ldots, 11$

r_1	n	1	2	3	4	5	6	7	8	9	10	11
0	$L_n(1)$	1	1	1	1	1	1	1	1	1	1	1
1	$L_n(5)$	1	3	4	7	11	18	29	47	76	123	199
2	$L_n(9)$	1	5	7	17	31	65	127	257	511	1025	2047
3	$L_n(13)$	1	7	10	31	61	154	337	799	1810	4207	9637
4	$L_n(17)$	1	9	13	49	101	297	701	1889	4693	12249	31021
5	$L_n(21)$	1	11	16	71	151	506	1261	3791	10096	29051	79531
6	$L_n(25)$	1	13	19	97	211	793	2059	6817	19171	60073	175099
7	$L_n(29)$	1	15	22	127	281	1170	3137	11327	33286	112575	345577

There is still more to be studied. The interested reader might like to extend some of these results to Golden Ratio Lucas numbers, $\{L_n(a)\}$, (Table 4.1.8), where it can be seen that they are related to the Golden Ratio Fibonacci numbers by

$$L_n(a) = F_{n+1}(a) + r_1 F_{n-1}(a) \tag{4.1.17}$$

which is analogous to the well-known (Subba Rao 1959).

$$L_n(5) = F_{n+1}(5) + F_{n-1}(5), \tag{4.1.18}$$

since $r_1 = 1$.

Other similar generalizations include

$$F_n(a) = \frac{1}{a}\left(L_{n+1}(a) + r_1 L_{n-1}(a)\right) \tag{4.1.19}$$

which is analogous to

$$F_n(5) = \frac{1}{5}\left(L_{n+1}(5) + L_{n-1}(5)\right). \tag{4.1.20}$$

Similarly

$$F_{m+n}(a) = \frac{1}{2}\left(F_m(a)L_n(a) + F_n(a)L_m(a)\right) \quad (4.1.21)$$

so that the generalizations are quite extensive. Further studies could also investigate these sequences in finite fields (Stein 1963), the graphs of their connections (Shannon and Horadam 1994), and the intersections between sequences with the same value for r_1 and among sequences with different values for r_1 (Horadam 1966; Stein 1962; Vajda 1989).

4.2. The Decimal String of the Golden Ratio[6]

It is well-known that the powers of the Golden Ratio, Phi or φ, are related to the elements of the Fibonacci sequence, $\{F_n\}$,(Atanassov et al. 2001):

$$\varphi^n = F_n\varphi + F_{n-1}, \quad (4.2.1)$$

and

$$\psi^n = F_{n+1} - F_n\phi, \quad (4.2.2)$$

in which

$$\psi = 1 - \varphi.$$

Similarly for the Lucas sequence, $\{L_n\}$:

$$\psi^n\Psi = L_{n-1}\psi + L_{n-2}, \quad (4.2.3)$$

[6] Based on Leyendekkers and Shannon (2014a).

and

$$\varphi^n \Phi = L_n - L_{n-1}\psi. \tag{4.2.4}$$

in which

$$\Psi = \psi + 2 \text{ and } \Phi = \varphi + 2$$

Are the roots of $0 = x^2 - 3x + 1$, which is the characteristic polynomial of the second order homogeneous linear recurrence relation

$$U_n = 3U_{n-1} - U_{n-2},$$

from which even- and odd-suffixed Fibonacci and Lucas numbers can be generated (Carlitz 1964); Hoggatt 1969). Variations of results for the Golden Section also include:

$$\lim_{n\to\infty} \frac{F_{n+6}}{F_n} = 8\varphi + 5, \tag{4.2.5}$$

and for $\{\varphi\}$, the decimal part of φ:

$$\begin{aligned} \{\phi\} &= \left(1 - \{\phi\}\right)^{\frac{1}{2}} \\ &= 1 - \sum_{r=1}^{\infty} C_r \left(F_r\phi - F_{r+1}\right) \\ &= \phi - 1, \end{aligned} \tag{4.2.6}$$

in which

$$C_r = \frac{\frac{1}{2}\left(-\frac{1}{2}\right)\ldots\left(\frac{3}{2} - r\right)}{r!},$$

so that

$$\varphi = \frac{\left(2+\sum_{r=1}^{\infty} C_r F_{r+1}\right)}{\left(1+\sum_{r=1}^{\infty} C_r F_r\right)}. \tag{4.2.7}$$

which may be compared with the known:

$$\varphi = \lim_{n\to\infty} \frac{F_{n+1}}{F_n}. \tag{4.2.8}$$

We note that (4.2.7) includes an irrational fraction on the right hand side, whereas (4.2.8) contains a rational fraction. φ may also be expressed as an expansion of the numerical value:

$$\begin{aligned}\varphi &= \tfrac{1}{2}\left(1+\sqrt{5}\right) \\ &= \tfrac{1}{2}+\sqrt{\tfrac{5}{4}} \\ &= \tfrac{1}{2}+(1+0.25)^{\frac{1}{2}} \\ &= 1.5+\sum_{r=1}^{\infty} C_r \left(\tfrac{1}{4}\right)^r.\end{aligned} \tag{4.2.9}$$

We now consider how these functions contribute to $\{\varphi\}$, the decimal string of φ. The progressive sums for Equations (4.2.6) and (4.2.9) are displayed in Table 4.2.1. For Equation (4.2.6) the first four decimal places are quickly achieved, but after the changes in the decimal places are relatively slow. As can be expected from such a direct calculation in Equation (4.2.9) which is not directly related to the Fibonacci numbers, the decimals are obtained almost sequentially with each r. The first seven decimal places are reached by $r = 9$.

Table 4.2.1. Comparisons of decimal expansions of φ

r	C_r	$\varphi = 1.5 + \sum_{r=1}^{\infty} C_r \left(\frac{1}{4}\right)^r$	$\varphi = 2 - \sum_{r=1}^{\infty} C_r \left(F_r\varphi - F_{r+1}\right)$
1	0.5000000	1.6250000	1.6909830
2	−0.1250000	1.6171875	1.6432385
3	0.0625000	1.6181640	1.6284842
4	−0.0390625	1.6180115	1.6227851
5	0.0273437	1.6180382	1.6205979
6	−0.0205078	1.6180332	1.6192004
7	0.0161132	1.6180341	1.6186448
8	−0.0130920	1.6180340	1.6183661
9	0.0109100	1.6180339	1.6182224
10	−0.0092735	1.6180339	1.6181232
11	0.0080089	1.6180339	1.6180838
…	…	…	…
15	0.0049815	…	1.61803999

The elements of the Lucas sequence also have:

$$\varphi = \lim_{n\to\infty} \frac{L_{n+1}}{L_n} \tag{4.2.10}$$

The first six decimal places for φ can be obtained from both F_{18}/F_{17} (Equation (4.2.8)) and L_{17}/L_{16}, and this also applies for $\left(\left(F_{23}/F_{17}\right)-5\right)/8$ from Equation (4.2.5).

If F_{n+1} is the length of a line subdivided into segments F_{n-1} and F_n, then, when

$$F_{n+1}/F_n \cong F_n/F_{n-1} \tag{4.2.11}$$

defines φ, that is

$$F_{n+1}F_{n-1} \cong F_n^2 \tag{4.2.12}$$

an approximation to Simson's identity, which is even when $n = 9$ yields

$$F_{n+1}F_{n-1} = 1155 \cong 1156 = F_n^2$$

and so the equality is approached more closely as n increases. However, as noted above in Equation (4.2.7), a ratio of irrationals gives a stronger result than a ratio of rationals. One can also infer this in another format by generalising the results in Table 6.5 of Havil (2012), namely,

$$\left|\varphi - \frac{F_{n+1}}{F_n}\right| < \frac{1}{F_n^2}.$$

For the ratios F_{n+1}/F_n and L_{n+1}/L_n, the values of n which correspond to the initial appearance of a decimal in the string for φ satisfy an inhomogeneous Pellian-like recurrence relation

$$n_{i+1} = 2n_i - n_{i-1} \pm 1 \tag{4.2.13}$$

as in Table 4.2.2.

Table 4.2.2. Values for *F*: Fibonacci; *L*: Lucas

Decimals	6	1	8	0	3	3	9	8	8
n_F	4	7	10	12	14	17	19	21	23
n_L	5	6	9	11	14	16	18	21	24

Bisection of sequences divides sequence in two. This can be further generalized with multisection (Carlitz 1964) of series which uses the primitive roots of unity to divide a given series into a number of sections. It is also related to lacunary recurrence relations (Shannon 1980) where there are gaps in the actual recurrence relation, such as.

$$F_n = 2F_{n-1} - F_{n-3}, \quad n \geq 4, \tag{4.2.14}$$

with initial conditions $F_1 = 1$, $F_2 = 1$, $F_3 = 2$, which generates the Fibonacci sequence. It is a third order equation but there are gaps between the two terms on the right hand side, but not within the sequence it generates.

We also note that the structure of the Fibonacci numbers, as analysed within the modular ring Z_5 for instance, prevents the approximation (4.2.12) from becoming an equality. The exact form is Simson's identity, namely

$$F_{n+1}F_{n-1} = F_n^2 + (-1)^n \tag{4.2.15}$$

For example,

$$F_{40}F_{38} = 4000054745112195$$

and

$$F_{39} = 4000054745112196,$$

which is also illustrated in Table 4.2.3 with the right-end-digits.

These results cn be extended to the whole golden secton family for further investigation by the interested reader.

Table 4.2.3. Right-end-digits ()* for Simson's identity

n	9	11	17	18	22	29	39	46
$(F_n^2)^*$	6	1	9	6	1	1	6	9
$(F_{n+1}F_{n-1})^*$	5	0	8	7	2	0	5	0

4.3. THE GOLDEN RATIO FAMILY AND THE BINET EQUATION

We have seen that the Golden Ratio may be considered as the first member of a family which can generate a set of generalized Fibonacci sequences. Here we relate the ideas there to the work of Filipponi (1991), Monzingo (1980) and Whitford (1977) to consider some related problems with their common thread being the Binet form of these sequences, $\{F_n(a)\}$, where the sequence of ordinary Fibonacci numbers can be expressed as $\{F_n(5)\}$ in this notation. Thus, for instance

$$\frac{F_n(a)}{F_{n-1}(a)} \to \varphi_a \tag{4.3.1}$$

in which

$$\varphi_a = \frac{1+\sqrt{a}}{2} \tag{4.3.2}$$

and the generalized Binet formula in this notation is

$$F_n(a) = \frac{\left(\frac{1+\sqrt{a}}{2}\right)^n - \left(\frac{1-\sqrt{a}}{2}\right)^n}{\sqrt{a}}. \tag{4.3.3}$$

which is well-known for the Fibonacci numbers as

$$F_n = \frac{\left(\frac{1+\sqrt{5}}{2}\right)^n - \left(\frac{1-\sqrt{5}}{2}\right)^n}{\sqrt{5}}. \tag{4.3.4}$$

Table 4.3.1. Classes and rows for $\mathbf{Z}_4$

Row $r_i\downarrow$	Class $i\rightarrow$	$\bar{0}_4$	$\bar{1}_4$	$\bar{2}_4$	$\bar{3}_4$	Comments
0		0	1	2	3	$N = 4r_i + i$
1		4	5	6	7	even $\bar{0}_4$, $\bar{2}_4$
2		8	9	10	11	$(N^n, N^{2n}) \in \bar{0}_4$
3		12	13	14	15	odd $\bar{1}_4$, $\bar{3}_4$; $N^{2n} \in \bar{1}_4$

Hence, elements of the sequences in the family should be similarly predicted. We note in passing that the Binet formula for the Fibonacci numbers is usually attributed to Jacques Philippe Marie Binet (1786–1856), but it was previously known to such famous mathematicians as Abraham de Moivre (1667–1754), Daniel Bernoulli (1700–1782), and Leonhard Euler (1707–1783): "like many results in Mathematics, it is often not the original discoverer who gets the glory of having their name attached to the result, but someone later!" (Knott 2015).

When n in Equation (4.3.3) is a power of 2 we can start to develop identities analogous to those of the Fibonacci sequence. For example,

$$x^{2n} - y^{2n} = (x^n - y^n)(x^n + y^n) \tag{4.3.5}$$

can become

$$F_{2n}(a) = F_n(a)L_n(a) \tag{4.3.6}$$

in which $L_n(a)$ is the corresponding generalized Lucas sequence. Both types of sequence satisfy the second order recurrence relation

$$u_n(a) = u_{n-1}(a) + r_1 u_{n-2}(a),\ n > 2, \tag{4.3.7}$$

where r_1 is in Class $\bar{1}_4 \in Z_4$ (a modular ring) (Table 4.3.1). We shall use this then with $r_1 = [(a-1)/4]$ as an integer in the recurrence relations which follow

We can continue the process in (4.3.5) to get

$$\begin{aligned} x^{2n} - y^{2n} &= \left(x^n + y^n\right)\left(x^n - y^n\right) \\ &= \left(x^n + y^n\right)\left(x^{\frac{n}{2}} + y^{\frac{n}{2}}\right)\left(x^{\frac{n}{2}} - y^{\frac{n}{2}}\right) \end{aligned} \tag{4.3.8}$$

and so on. For instance, when $n = 4$, this can be reduced to

$$x^8 - y^8 = \left(x^4 + y^4\right)\left(x^2 + y^2\right)(x+y)(x-y)$$

with $x + y = 1$ and $x - y = \sqrt{a}$, and when $n = 8$, this can be reduced to

$$x^{16} - y^{16} = \left(x^8 + y^8\right)\left(x^4 + y^4\right)\left(x^2 + y^2\right)(x+y)(x-y)$$

or

$$F_{16}(a) = L_8(a)L_4(a)L_2(a),$$

which can be readily confirmed when $a = 5$. More generally,

$$\frac{x^n - y^n}{x - y} = \left(x^{n-1} + y^{n-1}\right) + \frac{xy}{x-y}\left(x^{n-2} - y^{n-2}\right) \tag{4.3.9}$$

can be expressed as

$$F_n(a) = L_{n-1}(a) + \left(\frac{1-a}{4}\right)F_{n-2}(a), \tag{4.3.10}$$

which, when $a = 5$ and $n = 7$, $F_7(5) = 13$, and $L_6(5) - F_5(5) = 18 - 3$.

Equation (4.3.9) can be factorised further

$$\frac{x^n - y^n}{x - y} = \left(x^{n-1} + y^{n-1}\right) + \left(\frac{1-a}{4}\right)\left(x^{n-3} - y^{n-3}\right) + \left(\frac{1-a}{4}\right)^2\left(x^{n-5} - y^{n-5}\right) + \left(\frac{1-a}{4}\right)^3$$

This in turn can be re-written as

$$F_n(a) = L_{n-1}(a) + \left(\frac{1-a}{4}\right)L_{n-3}(a) + \left(\frac{1-a}{4}\right)^2 L_{n-5}(a) + \left(\frac{1-a}{4}\right)^3 \text{; (4.3.11)}$$

for instance,

$$F_7(5) = L_6(5) - L_4(5) + L_2(5) - 1.$$

Direct calculations of $\left(\frac{1+\sqrt{a}}{2}\right)^n$ and $\left(\frac{1-\sqrt{a}}{2}\right)^n$ as in the Binet equation (4.3.3) and from (4.3.10) yield the patterns set out in Table 4.3.2. Each n yields an infinity of 'golden ratios.'

That is, for example, as in (4.3.11):

$$\frac{1}{2^8}\left(a^4 + 36a^3 + +126a^2 + 84a + 9\right) \text{ so that}$$

$$u_9(5) = 34 = F_9(5),\ u_9(13) = 508 = F_9(13),$$

$$u_9(17) = 1165 = F_9(17),$$

in which the sequences satisfy the second order recurrence relation (4.3.7) in the form

$$u_n(a) = u_{n-1}(a) + \left(\frac{a-1}{4}\right)u_{n-2}(a),\ n > 2, \qquad (4.3.12)$$

with unity as the initial terms as in Whitford (1977). Thus Simson's identity becomes

$$F_n(a)F_{n+2}(a) - F_{n+1}^2(a) = (-1)^{n+1}\left(\frac{a-1}{4}\right)^n \qquad (4.3.13)$$

which had previously been proved by Lucas (1969). The work of Filipponi and Monzingo extended that of Whitford. In turn we can extend it further by considering the recurrence relation

$$v_n(a) = v_{n-1}(a) + 4(a-n)v_{n-2}(a),\ n > 2, \qquad (4.3.14)$$

which with unit initial conditions again generates the sequence

$$\{v_n(a)\} \equiv \{1,1,4a-11,8a-27,16a^2-120a+209,48a^2-412a+830,\ldots\},$$
$$\{v_n(0)\} \equiv \{1,1,-11,-27,193,\ldots\},$$
$$\{v_n(4)\} \equiv \{1,1,5,5,-15,\ldots\},$$
$$\{v_n(5)\} \equiv \{1,1,9,13,9,-30,\ldots\},$$

which invite a separate study particularly in relation to negative signs and intersections (Shannon 1983). Instead we shall briefly consider Whitford's table of sequences which we have slightly extended (Table 4.3.3).

We note that Table 4.3.3 is Sloane's A083856 with many individual rows and columns also listed there, as is the sequence of forward diagonals {1, 2, 3, 5, 9, 17, 34, 71, ...} [A110113] and the first backward diagonals {1, 3, 7, 29, 99, 463, ...} [A171180].

Table 4.3.2. Various Golden Ratio Sequences

n	$L_n(a)$	$L_n(5)$	$L_n(13)$	$L_n(17)$
2	$\frac{1}{2^1}(a+1)$	3	7	9
4	$\frac{1}{2^3}(a^2+6a+1)$	7	31	49
6	$\frac{1}{2^5}(a^3+15a^2+15a+1)$	18	154	297
8	$\frac{1}{2^7}(a^4+28a^3+70a^2+28a+1)$	47	799	1889
3	$\frac{1}{2^2}(a+3)$	2	4	5
5	$\frac{1}{2^4}(a^2+10a+5)$	5	19	29
7	$\frac{1}{2^6}(a^3+21a^2+35a+7)$	13	97	181
9	$\frac{1}{2^8}(a^4+36a^3+126a^2+84a+9)$	34	508	1165

Table 4.3.3. Whitford's table of Generalized Fibonacci numbers – extended

a	$\frac{a-1}{4}$	$F_1(a)$	$F_2(a)$	$F_3(a)$	$F_4(a)$	$F_5(a)$	$F_6(a)$	$F_7(a)$	$F_8(a)$	$F_9(a)$	$F_{10}(a)$
1	0	1	1	1	1	1	1	1	1	1	1
5	1	1	1	2	3	5	8	13	21	34	55
9	2	1	1	3	5	11	21	43	85	171	341
13	3	1	1	4	7	19	40	97	217	508	1159
17	4	1	1	5	9	29	65	181	441	1165	2929
21	5	1	1	6	11	41	96	301	781	2286	6191
25	6	1	1	7	13	55	133	463	1261	4039	11605

We choose now to consider finite difference operators, $\Delta_{i.j}$, (Rota et al. 1975), acting on the sequences within these rows and columns (i, j) for various values of a and n (the position of an element within each sequence $\{F_n(a)\}$). We define sequence row difference operators $\Delta_{a,j}F_j(a)$, sequence column difference operators $\Delta_{i,k}F_k(a)$, and sequence vector difference operators $\Delta_{a,k}F_k(a)$ (Shannon et al. 1991), respectively by

$$\Delta_{a,j}F_j(a) = F_j(a) - F_j(a-1) \tag{4.3.15}$$

and

$$\Delta_{i,k}F_k(i) = F_k(i) - F_{k-1}(i) \tag{4.3.16}$$

and

$$\Delta_{a,k}F_k(a) = F_k(a) - F_{k-1}(a-1) \tag{4.3.17}$$

We can then apply these operators to the rows, columns and forward and backward diagonals of the elements of Table 4.3.3 to find a variety of inter-related sequences, some expected, some unexpected, as we see in Tables 4.3.4, 4.3.5 and 4.3.6.

Table 4.3.4. First column differences, $\Delta_{i,k}F_k(a)$, from Table 4.3.3

$\Delta_{i,k}F_k(a)$ →	$F_2(a)$	$F_3(a)$	$F_4(a)$	$F_5(a)$	$F_6(a)$	$F_7(a)$	$F_8(a)$	$F_9(a)$	$F_{10}(a)$
$a = 1$	0	0	0	0	0	0	0	0	0
5	0	1	1	2	3	5	8	13	21
9	0	2	2	6	10	22	42	86	170
13	0	3	3	12	21	57	120	291	651
17	0	4	4	20	36	116	260	724	1764
21	0	5	5	30	55	205	480	1505	3905
25	0	6	6	42	78	330	798	2778	7566

Table 4.3.5. Second column differences, $\Delta_{i,k}(\Delta_{i,k}F_k(a))$, from Table 4.3.4

$\Delta^2_{i,k}F_k(a)$ →	$F_3(a)$	$F_4(a)$	$F_5(a)$	$F_6(a)$	$F_7(a)$	$F_8(a)$	$F_9(a)$	$F_{10}(a)$
$a=1$	0	0	0	0	0	0	0	0
5	1	0	1	1	2	3	5	8
9	2	0	4	4	12	20	44	84
13	3	0	9	9	36	63	171	360
17	4	0	16	16	80	144	464	1040
21	5	0	25	25	150	275	1025	2400
25	6	0	36	36	252	468	1980	4788

If we take the second five rows associated with the last six columns of Table 4.3.5 we get the patterns displayed in Table 4.3.6 in which the obvious common factor (the squares) in the fourth column of Table 4.3.5 has been taken out of each of the rows.

Table 4.3.6. Row patterns of second differences from Table 4.3.5

a	$\frac{a-1}{4}$	Common factor	$F_3(a)$	$F_4(a)$	$F_5(a)$	$F_6(a)$	$F_7(a)$	$F_8(a)$	OEIS number
5	1	1	1	1	2	3	5	8	A000045
9	2	4	1	1	3	5	11	21	A001045
13	3	9	1	1	4	7	19	40	A006130
17	4	16	1	1	5	9	29	65	A006131
21	5	25	1	1	6	11	41	96	A015440
25	6	36	1	1	7	13	55	133	A015441

4.4 SOME CHARACTERISTICS OF THE GOLDEN RATIO FAMILY[7]

The classical Golden Ratio, known since the time of Euclid (Livio 2002) is the first member of a family of surds, $\frac{1}{2}\sqrt{1+a}$, the members of

[7] Leyendekkers and Shannon (2016).

which have many similar properties], which we propose to explore further in this section. Other such research has included that of Kapur (1988) who used the fact that the Golden Ratio is related to the division of a line into two parts to then consider the division of a line into n parts as a way of generalising the Golden Ratio. Analogously, other generalizations have considered a 'silver mean' $\frac{2a+b}{a}=\frac{a}{b}$ related to the Pell numbers in a manner similar to the golden mean $\frac{a+b}{a}=\frac{a}{b}$, related to the Fibonacci numbers, and generalizations of the Fibonacci continued fraction (Shannon and Bernstein 1973). More recently, Crăciun *et al* (2015) have introduced a generalized Golden Ratio as a fixed point of an operator defined by an arbitrary mean, which satisfies certain conditions.In general as we saw in the last section, each member of the family satisfies

$$\varphi_a = \frac{1+\sqrt{a}}{2}, \qquad (4.4.1)$$

in which $a = 4r_1 + 1, a \in \bar{1}_4 \subset z_4$ (Table 4.4.1). The associated Fibonacci sequences satisfy (Berenhaut et al 2010):

$$F_{n+1} = F_n + r_1 F_{n-1}. \qquad (4.4.2)$$

Table 4.4.1. Classes and rows for Z_4

Row $r_i \downarrow$	Class $i \rightarrow$	$\bar{0}_4$	$\bar{1}_4$	$\bar{2}_4$	$\bar{3}_4$	Comments
0		0	1	2	3	• $N = 4r_i + i$
1		4	5	6	7	• even $\bar{0}_4, \bar{2}_4$
2		8	9	10	11	• $(N^n, N^{2n}) \in \bar{0}_4$
3		12	13	14	15	• odd $\bar{1}_4, \bar{3}_4$; $N^{2n} \in \bar{1}_4$

The Golden Ratio appears in the Elements of Euclid in several places (Livio 2002). Thus, if we take a line *AB* and divide by a point *C*, Euclid's definition of extreme and mean ratio is "(larger segment, *AC*) over (shorter segment, *CB*) is equal to (whole line, *AB*) over (larger segment, *AC*); that is,

$$AC/CB = AB/AC, \tag{4.4.3}$$

$$AC^2 = CB \times AB, \tag{4.4.4}$$

$$AB = AC + CB. \tag{4.4.5}$$

If $AC = \varphi_a$, then since

$$AC^2 = \varphi_a^2 = \varphi_a + r_1, \tag{4.4.6}$$

from Equations (4.4.3) to (4.4.6) we have:

$$\begin{aligned} AB^2 &= AB \times AC + AC^2 \\ &= AB \times \varphi_a + AC^2 \\ 0 &= AB^2 - \varphi_a AB - AC^2 \\ &= AB^2 - \varphi_a AB - \varphi_a^2 \end{aligned} \tag{4.4.7}$$

which has a solution

$$\begin{aligned} AB &= \tfrac{1}{2}\left(\varphi_a + \sqrt{\varphi_a^2 + 4\varphi_a^2}\right) \\ &= \tfrac{1}{2}\left(\varphi_a + \sqrt{5\varphi_a^2}\right) \\ &= \varphi_a\left(\frac{1+\sqrt{5}}{2}\right) \end{aligned}$$

$$AB = \varphi_a \varphi_5 \tag{4.4.8}$$

and

$$CB = \varphi_a (\varphi_5 - 1). \tag{2.7}$$

For example, when $a = 5$,

$$AB = \varphi_5^2 = \varphi_5 + 1. \tag{2.8}$$

We observe in Table 4.4.2 that the change in *AB* becomes increasingly smaller and *AB* reaches a constant value as $\varphi_{a+4} - \varphi_a$ approaches zero.

When $a = 5$, we have the well-known Simson's identity:

$$F_n^2 = F_{n+1} F_{n-1} + (-1)^{n-1}. \tag{4.4.9}$$

The question naturally arises does this identity extend to the whole Golden Ratio family? The answer is 'yes' in the form (Tables 4.4.3 and 4.4.4):

$$F_n^2 = F_{n+1} F_{n-1} + (-r_1)^{n-1}. \tag{4.4.10}$$

Table 4.4.2. *AB* and φ_a

a	φ_a	AB
5	1.6180339	2.6180339
9	2.0000000	3.2360618
13	2.3027756	3.7259689
17	2.5615528	4.1446876
21	2.7912878	4.5163982
25	3.0000000	4.8541017
29	3.1925824	5.1657065
33	3.3722813	5.4564654

Table 4.4.3. $a = 13$, $r_1 = 3$; $F_n^2 = F_{n+1}F_{n-1} + (-3)^{n-1}$.

n	F_n^2	F_{n+1}	F_{n-1}	$F_{n+1}F_{n-1}$	$(-r_1)^{n-1}$
2	1	4	1	4	–3
3	16	7	1	7	9
4	49	19	4	76	–27
5	361	40	7	280	81
6	1600	97	19	1843	–243
7	9409	217	40	8880	729

Table 4.4.4. $a = 17$, $r_1 = 4$; $F_n^2 = F_{n+1}F_{n-1} + (-4)^{n-1}$.

n	F_n^2	F_{n+1}	F_{n-1}	$F_{n+1}F_{n-1}$	$(-r_1)^{n-1}$
2	1	5	1	5	–4
3	25	9	1	9	16
4	81	29	5	145	–64
5	841	65	9	585	256
6	4225	181	29	5249	–1024
7	32761	441	65	28665	4096

The reciprocals of $(1 + \varphi_a)$ can be found from

$$R = \frac{r_1 - 2}{1 + \varphi_a} \qquad (4.4.11)$$

in which

$$R = \varphi_a - 2. \qquad (4.4.12)$$

Some examples are given in Table 4.4.4. Table 4.4.5 illustrates the consistency among the Golden Ratio family members.

Table 4.4.5. Reciprocals of $(1+\varphi_a)$

a	r_1	$(1+\varphi_a)$	$(1+\varphi_a)^{-1}$	$\frac{r_1-2}{1+\varphi_a}$
5	1	2.6180339	0.3819660	–0.3819660
13	3	3.3027756	0.3027756	0.3027756
17	4	3.5615528	0.2807764	0.5615528
21	5	3.7912878	0.2637626	0.7912878
29	7	4.1925824	0.2385164	1.1925820
33	8	4.3722813	0.2287134	1.3722804
37	9	4.5413812	0.2201973	1.5413811
41	10	4.7015621	0.2126952	1.7015621
45	11	4.8541010	0.2060113	1.8541017
53	13	5.1400544	0.1945504	2.1400544

Table 4.4.6. A 'golden circle'

a	$r = \sqrt{a} = 2\varphi_a - 1$	x	y
5	2.2360670	1	2
13	3.0605551	3	2
17	4.1231056	1	4
25	5.0000000	3	4
29	5.3851648	5	2
37	6.0827625	1	6

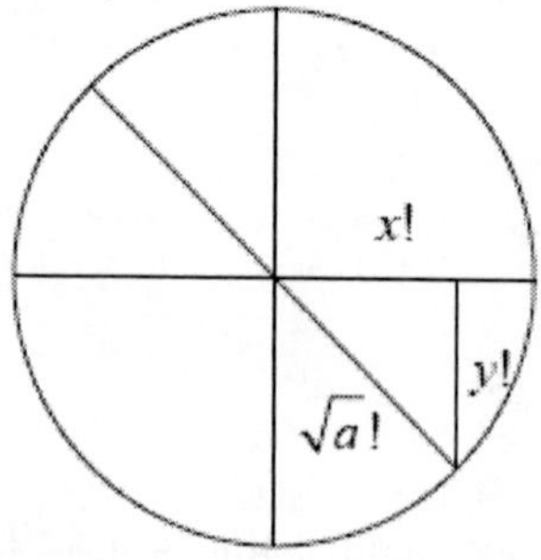

Figure 4.4.1. Equation (4.4.13).

Since $a \in \bar{1}_4, a$ can be a sum of squares (Table 4.4.6). Thus

$$a = x^2 + y^2 , \tag{4.4.13}$$

where *x*, *y* may be calculated from

$$x, y = \frac{A \pm \sqrt{2a - A^2}}{2} \tag{4.4.14}$$

with *x* odd and *y* even, and

$$A = x + y. \tag{4.4.15}$$

If a has factor in $\bar{3}_4$, then there could be no integers *x*, *y*. From Equations (4.4.13 and 4.4.14) $\sqrt{a}$ may be the radius of a circle in which *x* and *y* are integers as in Figure 4.4.1 and Table 4.4.6.

An infinity of circles with radius ($2\varphi_a - 1$), associated with the Golden Ratio, may thus be formed. This links φ_a to a variety of geometric figures such as cylinders, right circular cones, spheres and so on (Coxete 1953; Zaremba 1970).

4.5. THE COLLATZ CONJECTURE

In the (Lothar) Collatz Conjecture, odd integers are multiplied by 3 and 1 is added, whereas even integers are divided by 2. The sequential process is continued until unity is reached (Collatz 1986). The process has been called HOTPO: Half Or Triple Plus One and the property is called ‘oneness’. The elements of the Collatz sequence can be defined recursively:

$$c_0 = m,$$

$$c_n = \begin{cases} \frac{1}{2}c_{n-1}, & 2 \mid c_{n-1}, \\ 3c_{n-1}+1, & 2 \nmid c_{n-1}, \end{cases} \quad n > 0,\ c_n \neq 1. \tag{4.5.1}$$

The Collatz Conjecture is that process will which eventually reach 1 irrespective of the initial value.

$$\{C(m)\} \equiv \{m, c_1, c_2, \ldots, 1\} \tag{4.5.2}$$

Closely related to the Collatz Conjecture is the Boundedness Conjecture, namely that all Collatz sequences are bounded (Kaneda 2015). The number of integers in each Collatz sequence $C(m) \equiv \{c_n\}$ generated in this way is called the total stopping time (TST); that is, if the TST is i, then $c_i = 1$. The conjecture would be false if a sequence could be found with infinite TST. For instance, we see some finite cases in Table 4.5.1, in which we have disregarded 1,4,2,1 because of our restrictions in (4.5.1).

It can be seen that the powers of 2 converge quickly because 2^n is halved n times to reach one. Other even numbers produce an odd number, so that only initial odd numbers need to be considered.

Table 4.5.1. The first 8 Collatz sequences

m	$C_n = \{c_0, c_1, \ldots, c_n, \ldots, c_i\}$	TST
1	1	0
2	2,1	1
3	3,10,5,16,8,4,2,1	7
4	4,2,1	2
5	5,16,8,4,2,1	5
6	6,3,10,5,16,8,4,2,1	8
7	7,22,11,34,17,52,26,13,40,20,10,5,16,8,4,2,1	16
8	8,4,2,1	3

On the basis of this, investigations have been made of integers which contain powers of 2 such as the Mersenne and Fermat numbers, but neat patterns have failed to emerge. These and other specific approaches have led to a number of variations on the conjecture associated with the names of Helmut Hasse, Shizuo Kakutani, Brian Thwaites and Stanisław Ulam among others. A useful reference compendium is Lagarias (Lagarias 2010), which also has an inviting tree diagram on the cover and contains references to a virtual who's who of modern number theorists but excludes those who were wise enough to accept Guy's advice (Guy 1983), a paper which has been reproduced in the volume. The redoubtable Paul Erdös even said that "mathematics may not be ready for such problems" (Guy 2004). The purpose of this section is to analyse the conjecture in terms of modular rings and integer structure.

Variations on the original conjecture can be obtained with slight changes to (4.5.1). Conway (1972) proved that these Collatz-type problems are formally undecidable in a logical sense and that the original conjecture has no nontrivial cycles of length less than 400 in a computational sense. Lagarias later proved that there are no nontrivial cycles with length less than 275000 (Lagarias 1985).

Heuristic arguments and experimental evidence up to 5×10^{60} support the conjecture but since the integer structure remains constant to infinity it could be useful to examine the conjecture using the modular ring Z_4 (Table 4.5.2).

Table 4.5.2. Classes and rows for Z_4

Row $r_i \downarrow$	Class $i \rightarrow$	$\bar{0}_4$	$\bar{1}_4$	$\bar{2}_4$	$\bar{3}_4$	Comments
0		0	1	2	3	$N = 4r_i + i$
1		4	5	6	7	even $\bar{0}_4$, $\bar{2}_4$
2		8	9	10	11	$(N^n, N^{2n}) \in \bar{0}_4$
3		12	13	14	15	odd $\bar{1}_4$, $\bar{3}_4$; $N^{2n} \in \bar{1}_4$

The odd numbers in this ring fall in Classes $\bar{1}_4$ $(N = 4r_1 + 1)$ and $\bar{3}_4$ $(N = 4r_3 + 3)$, and the integers in these classes have distinctive features that make it critical to examine them separately. For example, in dealing with the golden ratio family $\varphi_a = \frac{1}{2}(1+\sqrt{a})$, a occurs only in Class $\bar{1}_4$.

Class $\bar{1}_4$ provides the components of all primitive Pythagorean triples because only integers in this class can equal a sum of squares. On the other hand Class $\bar{3}_4$ has more primes in regions where even powers abound. We now examine the reductions according to Collatz separately for these two classes.

Since

$$3N + 1 = 3(4r_1 + 1) + 1 = 4(3r_1 + 1) = 4r_0 \qquad (4.5.3)$$

the first reduction after class $\bar{1}_4$ is an even integer in class $\bar{0}_4$ (Table 2). When $3r_1 + 1$ is even (r_1 odd) the number can be reduced closer to 1, even though the Collatz numbers are often called the 'hailstone numbers' because their sometimes multiple ascents and descents are like hailstones in a cloud. Thus the class of N and the parity of the rows are important (Table 4.5.3).

Table 4.5.3. Class $\bar{1}_4$ and parity

Parity of r	Class of r_1	Class of ½(3N + 1)	Row of ½(3N + 1)
even	$4r_0$	$\bar{2}_4$	$6r_0$
	$4r_2 + 2$	$\bar{2}_4$	$6r_2 + 3$
odd	$4r_1 + 1$	$\bar{0}_4$	$6r_1 + 2$
	$4r_3 + 3$	$\bar{0}_4$	$6r_3 + 5$

Table 4.5.4. Structural Patterns

N	Row of N	Structural Patterns of the Class Sequences
5	$1\left(\bar{1}_4\right)$	100021
9	$2\left(\bar{2}_4\right)$	1023232102100210002 1
13	$3\left(\bar{3}_4\right)$	1002100021
17	$4\left(\bar{0}_4\right)$	1021002100021
21	$5\left(\bar{1}_4\right)$	10000021
25	$6\left(\bar{2}_4\right)$	1023210023210210021000 21
N	Row of N	Structural Patterns of the Class Sequences
29	$7\left(\bar{3}_4\right)$	1002321021002100021
33	$8\left(\bar{0}_4\right)$	10210232100232102100210002 1
37	$9\left(\bar{1}_4\right)$	10002323210210021000 21
41	$10\left(\bar{2}_4\right)$	1023232323210210212102323210232323210210023232 10232 10232323232323210021021002323232323232100000210021 0 0232102100210002 1
45	$11\left(\bar{3}_4\right)$	1002102100210002 1
49	$12\left(\bar{0}_4\right)$	10210002323210210021000 21
53	$13\left(\bar{1}_4\right)$	100002100021
57	$14\left(\bar{2}_4\right)$	1023210210210002323210210021000 21
61	$15\left(\bar{3}_4\right)$	1002323210000210002 1
65	$16\left(\bar{0}_4\right)$	102102100023232102100210002 1
69	$17\left(\bar{1}_4\right)$	100021002100021
73	$18\left(\bar{2}_4\right)$	10232321002323232102102323023232102323210210023232 1 02321023232323210021232323210232321023232102100232 1 02100232102100232100000023210023232102102321000230 2 32321002323232102102321002300002100232102100210002 1
105	$26\left(\bar{2}_4\right)$	10232323210021000232100232102100210002 1

Examples are set out in Table 4.5.4 in which the bars and subscripts which designate the classes are omitted for notational simplicity.

Since

$$3N+1=3(4r_3+3)+1=4(3r_1+2)+2 \tag{4.5.4}$$

the first conversion will be to class $\bar{2}_4$. The parity of r_3 determines whether division by 2 yields an integer in class $\bar{1}_4$ or $\bar{3}_4$ (Table 4.5.5).

Table 4.5.5. Class $\bar{3}_4$ and parity

Parity of r	Class of r_3	Class of ½(3N + 1)	Row of ½(3N + 1)
even	$4r_0$	$\bar{1}_4$	$6r_0 + 1$
	$4r_2 + 2$	$\bar{1}_4$	$6r_2 + 4$
odd	$4r_1 + 1$	$\bar{3}_4$	$6r_1 + 2$
	$4r_3 + 3$	$\bar{3}_4$	$6r_3 + 5$

Examples are set out in Table 4.5.6 in which the bars and subscripts which designate the classes are again omitted for notational simplicity.

The so-called hailstone effect can be seen in that the sequences can be very small or very large. Two examples of the latter are 63728127 which required 949 steps and 670617279 which required 986 steps. The constraints of the integer structure eventually leads to unity. These two integers are both in class $\bar{3}_4$ so that the first conversion is in class $\bar{2}_4$ then an odd integer.

There are sets of embedded sub-sequences which contain sequences that have already been shown to satisfy the conjecture. For example, when N = 41, the sequence is identical with that for N = 31; when N = 39, the sequence {32321023210002321002321021021002100021} contains the sub-sequence {321002321021002100021}, the sequence for N = 19. This, in turn, contains a sub-sequence of {32321021002100021}, the sequence for N = 7.

Table 4.5.6. Structural Patterns of the Class Sequences

N	Row of N	Structural Patterns of the Class Sequences
3	$0(\bar{0}_4)$	32100021
7	$1(\bar{1}_4)$	32321021002100021
11	$2(\bar{2}_4)$	321021002100021
15	$3(\bar{3}_4)$	3232321000002100021
19	$4(\bar{0}_4)$	321002321021002100021
23	$5(\bar{1}_4)$	32321000002100021
27	$6(\bar{2}_4)$	321023232323210210232102323210232323210210023232102321023 21023232323232102323232100021021021021000210023232100002100021
31	$7(\bar{3}_4)$	1023232323210210212102323210232323210210023232102321 023232323232321002102100232323232323210000021002100232102100210002 1
35	$8(\bar{0}_4)$	12100002100021
39	$9(\bar{1}_4)$	323210232100023210023210210210021000 21
43	$10(\bar{2}_4)$	3210210210002323210210021000 21
47	$11(\bar{3}_4)$	3232321021023212323210232323210210023232102321023232 32323210023232321000230210210232100002321023210002321023210021002100021
51	$12(\bar{0}_4)$	32100210023210210021002 1
55	$13(\bar{1}_4)$	3232100232323210210232102321023232321021002323210232 1023232323230023232323210002323232100023021021000210023232100002100021
59	$14(\bar{2}_4)$	3210232100023210023210210021000 21
63	$15(\bar{3}_4)$	32323232321000232102323210232323210210023 2 1023210000232323210232323232100021021021023210212100023210023210023210210021000 21
67	$16(\bar{0}_4)$	3210002321002321021002100021
71	$17(\bar{1}_4)$	32321021023210232321023232321021232321000000232100002100021
75	$18(\bar{2}_4)$	32102100000002 1

Table 4.5.7. Some imbedded sequences

N	Sequence	Structure
13	$1002\underline{100021}$	$1002C_5$
17	$102\underline{1002100021}$	$102C_{13}$
21	$1\underline{0000021}$	$1C_{16}$
25	$102\underline{3210023210210021000021}$	$102C_{19}$
29	$1002\underline{321021002100021}$	$1002C_{11}$
33	$102102\underline{3210023210210021000021}$	$102102C_{19}$
41	$102\underline{3232323210210212102323210232323210210002}$ $\underline{32321023210232323232323210021021002323 2}$ $\underline{323232321000002100210023210210021000 21}$	$102C_{31}$

Embedded sequences of this kind are also found for the extended Sophie Germain primes. For example, see Table 4.5.7 which is built mainly on the data in Tables 4.5.4 and 4.5.6.

Various authors have attempted neat "proofs" (Brookman 2008) and generalizations of the Collatz Conjecture of which one of the most interesting in experimental terms is that of Belaga and Mignotte (1998) (even though Conway has proved that a natural generalization of the conjecture is algorithmically undecidable).

The larger the number of steps the more the classes $\bar{2}_4$ and $\bar{3}_4$ occur. For $\bar{1}_4$ when N has a row in $\bar{2}_4$ the sequences are longer than for other new classes. The repeats of class $\bar{0}_4$ lead to more rapid reductions. Many of the sequences associated with the sequence may be found in the *On-line Encyclopedia of Integer Sequences* (Sloane 1973). If the size positions of N from Tables 4 and 5 are plotted against the corresponding TST then a spectral graph is obtained so that all N will have bounded TST; that is, the $3N$+1 and division by 2 will always lead to unity. Hence further studies could use a spectral-graph theory approach.

Chapter 5

TRANSCENDENTAL NUMBERS AND TRIANGLES

The Pascal–Fibonacci (PF) numbers for a given Fibonacci number sum to give the values of that Fibonacci number. Individual PF numbers are members of one of the triangular, tetrahedral or pentagonal series or have factors in common with the pyramidal or other geometric series. For composite numbers, partial sums of PF numbers can yield a factor, while prime Fibonacci numbers are detected with sums of squares.

Classes of the modular ring Z_4 can be substituted into convergent infinite series for π and $\sqrt{2}$ to obtain Q, the ratio of the arc of a circle to the side of an inscribed square to yield $\pi = 2\sqrt{2}\,Q$. The corresponding convergents of the continued fractions for π, $\sqrt{2}$ and Q are then considered, together with the class patterns of the modular rings $\{Z_4, Z_5, Z_6\}$ and decimal patterns for π.

The last section in this chapter defines an arbitrary order generalization of the Fibonacci and Lucas numbers. By defining the roots of the associated auxiliary equation in terms of the roots of unity, summation over compositions generates results for products of Fibonacci numbers related to Lucas numbers as well as integers in general. This then relates to the first section of Chapter Six.

5.1. THE PASCAL–FIBONACCI NUMBERS[8]

It is well-known that the numbers along the leading diagonals in the Pascal Triangle sum to numbers in the Fibonacci sequence $\{F_n\}$ (with generalizations to higher order recursive sequences and geometric dimensions (Shannon 1977). This provides a simple way to calculate the Fibonacci numbers without irrationals, as in the Binet equation; that is

$$F_n = 2 + \sum_{i=2}^{\frac{1}{2}(p-1)} \binom{p-i}{i-1}. \tag{5.1.1}$$

We have called elements of these Pascal–Fibonacci (PF) numbers, each of which is given by

$$N_{p_i} = \binom{p-i}{i-1} \tag{5.1.2}$$

which are listed in Table 5.1.1 for p from 7 to 59. For example, when $p = 17$ and $i = 4$, the third number in the sum is $N_{P_{17}}(3) = 286$. Similarly, when $p = 43$ and $i = 5$, $N_{P_{43}}(4) = 73815$. Again, when $p = 59$, the last $i = ½(p - 1) = 29$, so the 28th number in the sum is $N_{P_{59}}(28) = = 30!/28!\times 2! = 435$.

[8] Extracts from Leyendekkers and Shannon (2013).

Table 5.1.1. Pascal–Fibonacci numbers

p	F_p	
7	5,	6
11	9,	28,35,15
13	11,	45,84,70,21
17	15,	91,286,495,462,210,36
19	17,	120,455,1001,1287,924,330,45
23	21,	190,969,3060,6188,8008,6435,3003,715,66
29	27,	325,2300,10626,33649,74613,116280,125970,92378,43758,12376,1820, 105
31	29,	378,2925,14950,53130,134596,245157,319770,293930,184756,75582, 18564,2380,120
37	35,	561,5456,35960,169911,593775,1560780,3108105,4686825,5311735, 4457400,2704156,1144066,319770,54264,4845,171
41	39,	703,7770,58905,324632,1344904,4272048,10518300,20160075, 30045015,34597290,30421755,20058300,9657700,3268760,735471, 100947,7315,210
43	41,	780,9139,73815,435897,1947792,6724520,18156204,38567100, 64512240,84672315,86493225,67863915,40116600,17383860,5311735, 1081575,134596,8855231
47	45,	946,12341,111930,749398,3838380,15380937,48903492,124403620, 254186856,417225900,548354040,573166440,471435600,300540195, 145422675,51895935,13123110,2220075,230230,12650,276
53	51,	1225,18424,194580,1533939,9366819,45379620,177232627,563921995, 1471442973,3159461968,5586853480,8122425444,9669554100,936419 9760,7307872110,4537567650,2203961430,818809200,225792840, 44352165,5852925,475020,20475,351
59	57,	1540,26235,316251,2869685,20358520,115775100,536878650,2054455 634,6540715896,17417133617,38910617655,73006209045,1149558085 28,151532656696,166509721602,151584480450,113380261800,689232 64410,33578000610,12875774670,3796297200,834451800,131128140, 13884156,906192,31465,435

In this section, we examine the details of the structure of these numbers. In particular, the PF numbers are formed by sequential ratios of factorials and therefore have a regular structure. They are composed of primes, the maximum of which is P_{i-1} for F_{P_i}. For instance, the PF numbers associated with F_{19} have the primes {2, 3, 5, 7, 11, 13, 17} in varying proportions. The simple structure makes the position of each along the diagonals significant. For example, all the first numbers equal ($p - 2$), the second are triangular as are the last, while the third numbers are tetrahedral. The second last numbers, N_i, ($I = ½(p - 5)$), are pentagonal. We now look briefly at the triangular numbers which can be represented by

$$T_n = ½n(n + 1). \tag{5.1.3}$$

The second PF number of each F_p ($7 \leq p \leq 59$) are given by Equation (2.1) with $n = p - 4$ (Table 5.1.2).

In turn the tetrahedral numbers in this context are given by

$$H_n = \tfrac{1}{6}n(n+1)(n+2) \tag{5.1.4}$$

and the third PF numbers fit this series with $n = (p - 6)$ (Table 5.1.3).

Table 5.1.2. Second PF Numbers

p	n = p – 4	$N_2 = \frac{1}{2}n(n+1)$
7	3	6
11	7	28
13	9	45
17	13	91
19	15	120
23	19	190
29	25	325
31	27	378
37	33	561
41	37	703
43	39	780
47	43	946
53	49	1225
59	55	1540
61	57	1653

Table 5.1.3. Third PF Numbers

p	n = p – 6	$N_3 = \frac{1}{6}n(n+1)(n+2)$
11	5	35
13	7	84
17	11	286
19	13	455
23	17	969
29	23	2300
31	25	2925
37	31	5456
41	35	7770
43	37	9139
47	41	12341
53	47	18424
59	53	26235
61	55	29260

Table 5.1.4. $i = \frac{1}{2}(p-3)$

p	n = ½(p – 1)	Ni (Eq 2.1)	⅓Ni
7	3	6	2
11	5	15	5
13	6	21	7
17	8	36	12
19	9	36	15
23	11	66	22
29	14	105	35
31	15	120	40
37	18	171	57
41	20	210	70
43	21	231	77
47	23	276	92
53	26	351	117
59	29	435	145
61	30	465	155

Table 5.1.5. Pentagonal numbers $D_i = ½n(3n – 1)$ and position i

p	n = p2/24 *	Di = ½n(3n – 1)	position, i
11	5	35	3
13	7	70	4
17	12	210	6
19	15	330	7
23	22	715	9
29	35	1820	12
31	40	2380	13
37	57	4845	16
41	70	7315	18
43	77	8855	19
47	92	12650	21
53	117	20475	24
59	145	31465	24
61	155	35960	28

*residual of 0.04 for all *p* is neglected.

Assuming the pattern for *n* continues, that is $(p - 2i)$ for *n*, are the fourth numbers compatible with some geometric number series? Using $n = p - 8$ for the pyramidal numbers, that is

$$Q_n = \tfrac{1}{6}n(n+1)(2n+1) \tag{5.1.4}$$

it is found that these numbers always have a factor in common with the fourth PF numbers. In fact, the factor 5 is common to all fourth PF numbers except F_{19}, F_{29}, F_{59}; that is, when *p* has a right-end-digit (RED) of 9 and is therefore an element of the Class $\bar{4}_5 \subset Z_5$.

a) The last numbers are triangular numbers which satisfy Equation (5.1.3) with $n = ½(p – 1)$. All these numbers are divisible by 3 so that they fall into a special subset of the triangular numbers (Table 5.1.4).

b) The second last numbers N_i ($i = ½(p - 5)$) always have 5 as a factor. These numbers are always divisible by 5, and at least one of the factors is triangular with n varying from 1 to 22. Moreover, these numbers are pentagonal, given by:

$$D_n = \tfrac{1}{2}n(3n-1) \tag{5.1.6}$$

with $n = p^2/24$ (Table 5.1.5).

c) Third last numbers, N_i, $i = ½(p - 7)$, are always even (except for $p = 23$ or 41), and all are divisible by 7 (11 numbers from $p = 17$ to 59). Eight of the numbers also have 8 as a factor.

The remaining numbers have factors in common with triangular, tetrahedral and other geometric series, with $n = p - 2i$. They generally have a factor common to all, but there are a few exceptions: examples (Table 5.1.6):

Table 5.1.6. Remaining numbers, N_i

i	Common factor for i	Exceptions	
		p	Factor of Ni
4	5	29,59	3
5	3	13,19,31,37	7
6	7	41	11
7	5	31,41,47	3
8	3	53	11
9	5	29,59	11
10	11	43	3
11	5	29	7
12	5	31,37	7
13	5	37,53	7
14	5	59	11

Table 5.1.7. Partial sums of PF numbers

p	Fp	Numbers summed	Factors
19	4181	$S = N_1 + N_2 + N_3 = 17 + 120 + 145 = 37 \times 16$	37×113
37	24157817	$S = N_1 + N_2 = 35 + 561 = 4 \times 149$	$73 \times 149 \times 2221$

Table 5.1.8. Sums of squares

p	Fp	factors	d	e	d, e as Fn	
7	13	-	3	2	F4	F3
11	89	-	5	8	F5	F6
13	233	-	13	8	F7	F6
17	1597	-	21	34	F8	F9
19	4181	37,113	55	34	F10	F9
			41	50		
23	28657	-	89	144	F11	F12
29	514229	-	377	610	F14	F15
31	1346269	557,2417	987	610	F16	F15
			875	762		
37	24157817	73,149,2221	4181	2584	F19	F18
			4909	244		
			3859	3044		

We have seen in the last chapter how the structure of the Fibonacci numbers can help to identify primes. The PF numbers can be partially summed to find a suitable factor (Table 5.1.7).

Another approach is via the sum of squares:

$$F_p = d^2 + e^2. \tag{5.1.7}$$

Primes only have one set of (d, e) with no common factors. Generally composites have the same number of sets as their factors. One (d, e) set is given by

$$F_p = \left(F_{\frac{1}{2}(p+1)}\right)^2 + \left(F_{\frac{1}{2}(p-1)}\right)^2 \tag{5.1.7}$$

Examples are displayed in Table 5.1.8.

A simple algebraic form to calculate (x,y) couples may also be used as in Table 5.1.9; that is,

$$x, y = \frac{1}{2}\left(A + \sqrt{2F_p - A^2}\right).$$

Table 5.1.9. Calculated (*x*,*y*) couples

p	Fp	$\sqrt{2F_p}$	A	d	e
7	13	5.1	5	3	2
11	89	13.3	13	5	8
13	233	21.6	21	13	8
17	1597	56.5	55	21	34
19	4181	91.4	89,91	55	34
				41	50
23	28657	239.4	233	89	144
29	514229	1014.1	987	377	610
31	1346269	1640.9	1637,1597	987	610
				875	762
37	24157817	6950.9	6903,6765,5153	9181	2584
				9909	3544
				3859	3044

Table 5.1.10. $(x^2 + y^2)^*$

	x*	1	3	5	7	9
	(x2)*⇒	1	9	5	9	1
y*	(y2)*⇓					
0	0	1	9	5	9	1
2	4	5	3	9	3	
4	6	7	5	1	5	
6	6	7	5	1	5	
8	4	5	3	9	3	

From Table 5.1.10, for $(x^2 + y^2)^* = 9$, only $((x^2)^*, (y^2)^*=$ (3,0); (5,8); or (5,2) will yield 9 so that $A^* = 3$ or 7.

5.2. The Structure of 'Pi'

The emphasis in this book has been on Fibonacci numbers and integer structure, it is of interest to note that the structure of the transcendatal number 'pi' is pertinent because of the result connecting pi with the arctangents of the reciprocals of alternate odd-indexed Fibonacci numbers (Lehmer 1936).

The number 'Pi', π, has been studied since antiquity (Ball and Coxeter 1956). In particular, various mathematicians have developed calculations around inverse tangents and power series: John Wallis (1616–1703) [see Equation (5.2.1)], James Gregory (1638–1675) Shanks and Wrench 1962), Gottfried Wilfred Leibnitz (1646–1716) (Nimbran 2010), John Machin (1680–1751) (Todd 1949), Leonard Euler (1701–1783) (Bennett 1925), Robert Simson (1687–1768) (Birch 1946), Carl Friedrich Gauss (1777–1855) (Sierpinski 1964), to mention but a few famous names from the history of mathematics.

With the dawn of differential calculus, the Greek method of inscribed and circumscribed polygons was replaced by convergent infinite series and algebraic and trigonometric methods. Wallis' elegant formula was

$$\frac{\pi}{2} = \frac{2}{1} \times \frac{2}{3} \times \frac{4}{3} \times \frac{4}{5} \times \frac{6}{5} \times \frac{6}{7} \times \frac{8}{7} \times \frac{8}{9} \times \ldots \tag{5.2.1}$$

Johan Heinrich Lambert (1728–1777) was the first to provide a rigorous proof that π is incommensurable (Kasner and Newman 1959): π cannot be the root of a rational algebraic equation. The advent of electronic computers has since revived interest in a variety of techniques which have also enriched pure mathematics (Lehmer 1938; Todd 1949): Google now has more than 1 million places listed! What more can be said? Here we outline how Integer Structure Analysis (ISA) permits a different analysis of the infinite series for π through functions of the rows of modular rings when detailed as modular arrays as in Table 5.2.1.

Table 5.2.1. Rows of Z_4

Row	f(r)	$4r_0$	$4r_1+1$	$4r_2+2$	$4r_3+3$
	Class	$\overline{0}_4$	$\overline{1}_4$	$\overline{2}_4$	$\overline{3}_4$
0		0	1	2	3
1		4	5	6	7
2		8	9	10	11
3		12	13	14	15
4		16	17	18	19
5		20	21	22	23
6		24	25	26	27
7		28	29	30	31

In a circle of radius *r*, the arc length of a quadrant is $\frac{1}{2}\pi r$, and the length of the side of an inscribed square is $\sqrt{2}r$, so that the ratio of the arc of length of the quadrant to the side of the inscribed square is $\pi/2\sqrt{2}$. We now use the modular ring Z_4 to convert Equation (5.2.1) to

$$\frac{\pi}{2}=\prod_{r=0}^{\infty}\frac{(4r+2)^2(4r+4)^2}{(4r+1)(4r+3)^2(4r+5)} \tag{5.2.2}$$

combining segments of four fractions for each *r*, and

$$\sqrt{2}=\prod_{r=0}^{\infty}\frac{(4r+2)^2}{(4r+2)^2-1} \tag{5.2.3}$$

so that from (5.2.2) and (5.2.3) we have that

$$\frac{\pi}{2\sqrt{2}} = \prod_{r=0}^{\infty} \frac{(4r+4)^2}{(4r+4)^2 - 1} \tag{5.2.4}$$
$$= Q.$$

If we use the values of π and $\sqrt{2}$ to 30 decimal places, we arrive at

$Q = 1.110720734882156630963 7,$

and when $r = 175$, Equation (5.2.4) yields

$Q = 1.1107212.$

The series for Q and $\sqrt{2}$ are similar in format: the product of even and odd fractions. The numerator, $4r + 2$, for $\sqrt{2}$ is in Class $\bar{2}_4$. Thus, the even numerator of the multiple fractions produced will contain odd factors for all values of r, whereas the variable $(4r + 4) \in \bar{0}_4$ will not, and those factors which do occur will cancel out. So the structures of the product fractions are

$$\frac{2^n p_1 p_2 \cdots}{p_1' p_2'} \text{ for } \sqrt{2}$$

and

$$\frac{2^m}{p_1 p_2 \cdots} \text{ for } Q.$$

The structure of π is thus resolved into

$$\pi = 2\sqrt{2}Q. \tag{5.2.5}$$

Q is irrational but can be useful in the analysis of the decimal expansion of π obtained from Q and $2\sqrt{2}$. Moreover, the "factorisation" of π should be useful in the context of continued fractions as we now illustrate. Every positive number can be expressed uniquely as a regular continued fraction (Kasner and Newman 1959).

$$a_0 + \cfrac{1}{a_1 + \cfrac{1}{a_2 + \cfrac{1}{a_3 + \ldots}}} = a_0 + \frac{1}{a_1 +}\frac{1}{a_2 +}\frac{1}{a_3 +}\cdots \tag{5.2.6}$$
$$= [a_0; a_1, a_2, a_3, \ldots].$$

The partial quotients $\{a_1, a_2, a_3, \ldots\}$ for $\sqrt{2}$ have a simple pattern

$$\sqrt{2} = [1; 2,2,2,2,2,\ldots],$$

but no such pattern has been found for π.

Convergents have remarkable properties and they are obtained by stopping the continued fractions at each successive stage (Mack 1970). The first six convergents for $\sqrt{2}$ and π may be used to calculate those for Q (Table 5.2.2).

Table 5.2.2. First six convergents for $\sqrt{2}$, π and Q

Number	Convergents					
$\sqrt{2}$	$\frac{1}{1}$	$\frac{3}{2}$	$\frac{7}{5}$	$\frac{17}{12}$	$\frac{41}{29}$	$\frac{99}{70}$
π	$\frac{3}{1}$	$\frac{22}{7}$	$\frac{333}{106}$	$\frac{355}{113}$	$\frac{103993}{33102}$	$\frac{104348}{33215}$
Q	$\frac{3}{2}$	$\frac{22}{21}$	$\frac{1665}{1484}$	$\frac{2130}{1921}$	$\frac{3015797}{2714364}$	$\frac{104348}{93951}$

The convergents $\frac{b}{c}$ satisfy the second order linear recurrence relations (Nimbran 2010):

$$b_{n+1} = a_n b_n + b_{n-1},\ c_{n+1} = a_n c_n + c_{n-1}. \tag{5.2.7}$$

with suitable initial conditions. Thus, in the first row of Table 5.2.2, the numerators and denominators of the convergent to $\sqrt{2}$ satisfy the Pell recurrence relation in which $a_n = 2$, with initial terms 1 and 3 in the numerators $\{b_n\}$, and 1 and 2 in the denominators $\{c_n\}$ (the standard Pell sequence (Horadam 1971)). There too $c_{n-1} + c_n = b_n$, analogous to the identity which relates the Fibonacci and Lucas numbers. We further observe that if $u_n = b^2 c^2$ where $\frac{b}{c}$ is any convergent to the continued fraction for $\sqrt{2}$, then the elements of

$$\{u_n\} = \{1,36,1225,41616,1413721,48024900,..\}$$

are square-triangular numbers (Lafer 1971) with the formal generating function

$$\sum_{n=0}^{\infty} u_n x^n = \frac{1+x}{1-x+x^2-x^3}. \tag{5.2.8}$$

Questions about square-pentagonal and triangular-pentagonal numbers are still open. Finally, we note that $Q = \frac{1}{2}\left(\sqrt{1.4\dot{9}\dot{1}} + 1\right)$, in which the argument of the radical is rational. The consecutive positions of the decimals of π do no appear to have a regular pattern, but we can analyse these positions indirectly. We do these by classifying them in terms of their position, n, in the array of a modular ring. See Tables 5.2.1, 5.2.3, 5.2.4 for Z_4, Z_5 and Z_6.

Table 5.2.3. Rows of Z_5

Row	f(r)	$5r_0$	$5r_1+1$	$5r_2+2$	$5r_3+3$	$5r_4+4$
	Class	$\bar{0}_5$	$\bar{1}_5$	$\bar{2}_5$	$\bar{3}_5$	$\bar{4}_5$
0		0	1	2	3	4
1		5	6	7	8	9
2		10	11	12	13	14
3		15	16	17	18	19
4		20	21	22	23	24
5		25	26	27	28	29

Table 5.2.4. Rows of Z_6

Row	f(r)	$6r_1-2$	$6r_2-1$	$6r_3$	$6r_4+1$	$6r_5+2$	$6r_6+3$
	Class	$\bar{1}_6$	$\bar{2}_6$	$\bar{3}_6$	$\bar{4}_6$	$\bar{5}_6$	$\bar{6}_6$
0		-2	-1	0	1	2	3
1		4	5	6	7	8	9
2		10	11	12	13	14	15
3		16	17	18	19	20	21
4		22	23	24	25	26	27
5		28	29	30	31	32	33

The sequence of classes provides patterns that have some repetitions for the 100 decimal places considered (Table 5.2.5), but do these patterns persist for the more than billion places already known?

If each 100 decimals is placed in a 10×10 array, the column, row and number n of the position of the decimal in the array will be characteristic of that number. Comparison of sequential 100 decimals can then be made. n^*, the right-end-digit (RED) of n, will equal the column in which the number falls. For example, if $n = 9, 19, 29, 39,\ldots$, the number will fall in column 9 (Tables 5.2.6).

Table 5.2.5. Class patterns for *n*
(position of number in decimal array for π)

	o e	Total	Z4	Z5	Z6
1	5 3	8	13301023	13204340	46414512
2	8 4	12	201011310311	111333331343	316162641226
3	7 4	11	13101332023	40240231411	66234241154
4	5 5	10	2330130230	2431240022	5423623165
5	3 5	8	00230312	43013110	15143643
6	4 5	9	302110322	202142023	451263615
7	5 3	8	11330203	34421114	42625336
8	4 8	12	322230322100	131402243143	235121453631
9	4 10	14	10222201322300	02403240032400	23535356415451
0	5 3	8	022113111	20401202	55322244

Legend: o=odd; e=even; bars and subscripts are omitted for notational brevity in the elements of Z4,Z5,Z6: e.g., $\bar{3}_4$ is represented by 3, etc.

When 300 decimal places are considered (three arrays), each decimal number has characteristic features in terms of rows in which they do not occur (Table 5.2.7). Also the appearances of these numbers in the sequence, n, (1st decimal $n = 1$) have certain REDs that do not occur. For instance, for 7, n never has a RED of 1, while 3 never has a RED of 8. Thus for the first 300 decimals, 7 never occurs for $n = 1, 11, 21, 31, \ldots, 291$, while 3 never occurs for $n = 8, 18, 28, 38, \ldots, 298$.

Billions of decimal places have been calculated for π and it continues to be the subject of many papers. REDs and ISA with modular rings introduce some new perspectives.

Finally, it is of interest to note that from the work of the famous computational mathematician, Derek Lehmer (Lehmer 1936), came the neat and relevant result, referred to at the start of this section, namely that the arctangents of the reciprocals of alternate odd-indexed Fibonacci numbers sum to $\pi/4$.

Finally, we note that the form of Q following Equation 5.2.8 resembles that of the generalized Golden Ratio Family, Φ_a.

Table 5.2.6. (a) $N = 2^n$

N = 2			N = 4			N = 8		
Col	Row	n	Col	Row	n	Col	Row	n
6	1	6	2	1	2	1	2	11
6	2	16	9	2	19	8	2	18
1	3	21	3	3	23	6	3	26
8	3	28	6	4	36	4	4	34
3	4	33	7	6	57	5	4	35
3	6	53	9	6	59	2	6	52
3	7	63	10	6	60	7	7	67
3	8	73	10	7	70	4	8	74
6	8	76	7	9	87	8	8	78
3	9	83	2	10	92	1	9	81
3	10	93				4	9	84
						8	9	88

Table 5.2.6(b). $3|N$

N = 3			N = 6			N = 9		
Col	Row	n	Col	Row	n	Col	Row	n
9	1	9	7	1	7	5	1	5
5	2	15	10	2	20	2	2	12
7	2	17	2	3	22	4	2	14
4	3	24	1	5	41	10	3	30
5	3	25	9	7	69	8	4	38
7	3	27	2	8	72	2	5	42
3	5	43	5	8	75	4	5	44
6	5	46	2	9	82	5	5	45
4	7	64	8	10	98	5	6	55
6	9	86				8	6	58
1	10	91				2	7	62
						9	8	79
						10	8	80
						10	10	100

Table 5.2.6(c). $N < 9$

N = 0			N = 1			N = 5			N = 7		
Col	Row	n	Col	Row	n	Col	Row	n	Col	Row	n
22	4	32	1	1	1	4	1	4	3	2	13
10	5	50	3	1	3	8	1	8	9	3	29
4	6	54	7	4	37	10	1	10	9	4	39
5	7	65	10	4	40	1	4	31	7	5	47
1	8	71	9	5	49	8	5	48	6	6	56
7	8	77	8	7	68	1	6	51	6	7	66
5	9	85	4	10	94	1	7	61	6	10	96
7	10	97				10	9	90			

Table 5.2.7. Distribution of decimals $n = 1–300$

Rows missing					n* missing			
Decimal Numbers	1st 100	2nd 100	3rd 100	1–300	1st 100	2nd 100	3rd 100	1–300
0	1,2,3	9	2,3,4,6,8	---	3,6,8,9	0,3	1,2,3,6,9	3
1	2,3,6,8,9	2,3,9	3,6,8	3	2,5,6	1,2,6,7,9	4,2	2
2	5	3,10	2,4,7	---	0,2,4,5,7,9	1,7,8	2,6,7	7
3	4,6,8	1,6,8,9	1,5,6	6	0,2,8	0,2,8,9	8,9	8
4	5,8	4	4,5,9	---	1,4,5,8	0	5,9	---
5	2,3,8,10	2,7,9	5,7,9,10	---	2,3,5,6,7,9	4,5,6	1,4,7,8,9	---
6	4,6	4,5,6,7,8	1,5,10	---	3,4,6	2,3,5,6,9	5	---
7	1,8,9	1,3,5,8,9,10	2,6,7,8	8	0,1,2,4,5,8	1,3,5,8,9	1,3,6,7,10	1
8	1,5,10	---	2,5,9,10	---	3,9	6,8	1,10	---
9	9	1,2,4,6	3,4,7,8	---	1,3,6,7	1,5,6,7,8	1,2,3,5,6,10	1,6

5.3. PELLIAN SEQUENCE RELATIONSHIPS AMONG Π, E, $\sqrt{2}$ [9]

From the previous section we have that

[9] Extracted from Leyendekkers and Shannon (2012).

$$\pi = 2\sqrt{2}\,Q \tag{5.3.1}$$

where Q is the ratio of the quarter circumference of a circle to the side of the inscribed square in Section 5.2. Here we extend the study to the structure of the irrationals e and $\sqrt{2}$ and compare with π. The first six convergents of their continued fractions are set out in Table 5.2.1.

The convergents $\frac{N}{D}$ from the first row of Table 5.3.1 satisfy the second order linear recurrence relations (Horadam and Shannon 1988):

$$N_{n+1} = 2N_n + N_{n-1},\ D_{n+1} = 2D_n + D_{n-1}. \tag{5.3.2}$$

with initial terms 1 and 3 in the numerators $\{N_n\}$, and 1 and 2 in the denominators $\{D_n\}$ (the standard Pell sequence (Horadam and Shannon 1993).

From the relationship between the Pell and Pell-Lucas sequences, it has been shown (Leyendekkers and Rybak 1995b) that Pellian sequences can be generated from the z–j grid (Leyendekkers and Rybak 1995a) set up to characterise Pythagorean triples (Table 5.3.2).

$$c^2 = b^2 + a^2,\ b > a$$

Two questions immediately arise:

- Are the sequences, $\{N_n\}$ and $\{D_n\}$ related to primitive Pythagorean triples (pPts)?, and
- Are there similar structures for π and e?

The elements of the numerator sequence, $\{N_n\}$, are all odd and it is found that they equal d $(= j)$ for primitive Pythagorean triples with $z = 1$ (Table 5.3.3). The sequences $\{f\}$ and $\{y\}$ seem to be new. However, $\{f\} = \{2u_n\}$ where

$$u_{n+1} = 2u_n + u_{n-1} - 1.$$

a Pellian non-homogeneous second order recurrence relation with initial terms, 1 and 2. That is, {*f*} satisfies

$$f_{n+1} = 2f_n + f_{n-1} - 2.$$

Table 5.3.1. First six convergents for $\sqrt{2}$, π and *e*

Number	Convergents					
$\sqrt{2}$	$\frac{1}{1}$	$\frac{3}{2}$	$\frac{7}{5}$	$\frac{17}{12}$	$\frac{41}{29}$	$\frac{99}{70}$
π	$\frac{3}{1}$	$\frac{22}{7}$	$\frac{333}{106}$	$\frac{355}{113}$	$\frac{103993}{33102}$	$\frac{104348}{33215}$
e	$\frac{2}{1}$	$\frac{3}{1}$	$\frac{8}{3}$	$\frac{11}{4}$	$\frac{19}{7}$	$\frac{87}{32}$

Table 5.3.2. *z–j* grid for Pythagorean triples: *j* is the integer counter; criterion for generating pPts is (*j*, *z*½) = 1 when *z* > 1; if *z* = 1 only pPts are obtained

z = c – b, b > a		c	b	a	y = b – a
odd	(2K – 1)2	j2 + (j + z½)2 d2 + f2	2j(j + z½) 2df	z½(2j + z½) f2 – d2	2j2 – z
even	2K2	[($\frac{1}{2}$ z)½ + 2j – 1]2 + $\frac{1}{2}$ z d2 + f2	[($\frac{1}{2}$ z)½ + 2j – 1]2 – $\frac{1}{2}$ z f2 – d2	[($\frac{1}{2}$ z)½+2j–1] × 2($\frac{1}{2}$ z)½ 2df	(2j – 1)2 – z

Table 5.3.3. Numerators and pPts

n	d = j	f = j + z½	pPts	y = 2j2 – z	z = 2j2 – y
1	1	2	5, 4, 3	1	1
2	3	4	25, 24, 7	17	1
3	7	8	113, 112, 15	97	1
4	17	18	613, 612, 35	577	1
5	41	42	3445, 3444, 83	3361	1
6	99	100	19801, 19800, 199	19601	1
7	239	240	114721, 114720, 479	114241	1
8	577	578	667013, 667012, 1155	665857	1

It should be noted that all the components of the major component of the pPts in both Tables 5.3.3. and 5.3.5 have their origins in Class $\bar{1}_4 \subset Z_4$ and thus have the form $4r_1 + 1$ where r_1 is the row. using simple algebra d and f can be calculated from

$$d, f = \tfrac{1}{2}\left(A + \sqrt{2c - A^2}\right)$$

with d odd and f even (Table 5.3.4).

Table 5.3.4. Calculation of *d* and *f*

Row, rl	c = 4rl + 1	√2c	A	d,f
1	5	3.16	3	(1,2)
6	25	7.07	7	(3,4)
28	11	15.03	15	(7,8)
153	613	35.01	35	(17,18)
861	3445	83.00	83	(41,42)
4950	19801	199.00	199	(99,100)
28680	114721	479.00	479	(239,240)
166754	667013	1155.00	1155	577,578)

The other internal parameters are z = 1, and y (c – b) which also satisfies a Pellian non-homogeneous recurrence relation:

$$y_{n+1} = 6y_n - y_{n-1} + a \tag{5.3.3}$$

in which

$$a = \begin{cases} -4 & n \text{ even,} \\ 12 & n \text{ odd.} \end{cases}$$

The elements of the sequence of denominators, $\{D_n\}$, equal d, f pairs for pPts with y (b – a) = 1 (Table 5.3.5).

Again we can find Pellian-type recurrence relations; for instance, $\{\sqrt{z}\}$ satisfies (5.3.3) with a = 0. The convergents of e in the third row of Table 5.3.1 oscillate between Pellian and Fibonacci sequences (Table 5.3.6).

The convergents of π in the second row of Table 1 also oscillate between Pellian and Fibonacci sequences (Table 5.3.7).

Table 5.3.5. Denominators and pPts

n	d = j	f = j + z½	pPts	z	$\sqrt{z}$	y
1, 2	1	2	5, 4, 3	1	1	1
3, 4	5	12	169, 120, 119	49	7	1
5, 6	29	70	5741, 4060, 4059	1681	41	1
7, 8	169	408	195025, 137904, 137903	57121	239	1

Table 5.3.6. Recurrence relations for convergents of *e*

n	1,2,3	2,3,4,5	4,5,6
Recurrence relation	$N_n = 2N_{n-1} + N_{n-2}$	$N_n = N_{n-1} + N_{n-2}$	$N_n = 4N_{n-1} + N_{n-2}$
Type	Pellian	Fibonacci	Pellian

Table 5.3.7. Recurrence relations (Rr) for convergents of π

n	1,2,3	2,3,4	3,4,5	4,5,6
Recurrence relation	$N_n = 15N_{n-1} + N_{n-2}$	$N_n = N_{n-1} + N_{n-2}$	$N_n = 292N_{n-1} + N_{n-2}$	$N_n = N_{n-1} + N_{n-2}$
Type	Pellian	Fibonacci	Pellian	Fibonacci

The Pellian-type sequences are again associated with pPts. For example, for π the first two N_n are the d and f of the triple: {493, 475, 132} with z = 18 and y = 343. If N_3 (333) is taken as d, then the triple is {757305, 535527, 535464} with z = 221778 (18 × 12321) and y = 63. When D_n = 7, 106 for d and f, this yields the triple {11285, 11187, 1484} with z = 98 and y = 9703. The value of z is even and has a right-end-digit (RED – sometimes termed "the digital root of an integer" (Trigg 1980) of 8 while y has a RED of 3. The REDs remain constant while the values of z and y vary. This is in contrast to the $\sqrt{2}$ system where $z = 1$ for $\{N_n\}$ and $y = 1$ for $\{D_n\}$.

The continued fraction for π is:

$$\pi = 3 + \frac{1}{7+}\frac{1}{15+}\frac{1}{1+}\frac{1}{292+}\frac{1}{1+}\frac{1}{1+}\frac{1}{1+}\frac{1}{2+}\frac{1}{1+}\frac{1}{3+}\frac{1}{1+}\cdots \qquad (5.3.4)$$

The occurrence of 15, 1 and 292, 1 and 2, 1 in the partial quotients and the coefficients in the Pellian-type recurrence relations invites further investigation. These partial quotients, unlike those for e and $\sqrt{2}$, have not been found to obey any simple laws (Ball and Coxeter 1956). It is somewhat surprising then that the Pellian relations for e and π have similar patterns which are a mix of the second order recurrence relations (Shanks and Wrench 1962).

The recurrence relations associated with $\sqrt{2}$ follow different patterns in their links with pPts, in contrast to those of e and π.

Table 5.3.8. Examples of recursive sequence defined by (5.3.6)
To what extent can these results be generalized?

m \ n	1	2	3	4	5	6	Reference
1	1	1	5	29	169	985	Fielder (1968)
2	1	2	11	64	373	2174	Emerson (1969)
3	1	3	17	99	577	3363	Lehmer (1935)
4	1	4	23	134	781	4552	Adler (1969)
5	1	5	29	169	985	5741	Fielder (1968)
6	1	6	35	204	1189	6930	Forder (1963)
7	1	7	41	239	1393	8119	Newman, Shanks, Williams (1980/81)
Δ_2	0	1	6	35	204	1189	Forder (1963)

Topics for further research readily emerge. For instance, if we take the recurrence relation (5.3.3) and generalize it to the homogeneous form

$$w_{m,n} = 6w_{m,n-1} - w_{m,n-2} \tag{5.3.5}$$

with initial conditions $w_{m,1} = 1, w_{m,2} = m$, $m = 1,2,\ldots,7$, we get the tableau in Table 5.3.8.

5.4. Extensions to the Zeckendorf Triangle

This section extends a result in Shannon (2010) which built on some work by Griffiths (2010) on a form of the Zeckendorf Triangle. A left-corrected form appears in Table 5.4.1 where it can be seen that the columns, $\{F_{m,n}\}$, are Fibonacci number multiples, $F_{m,n} = F_{m-n+1}F_n$, $m > n$, of the numbers in the Fibonacci sequence, in which m and n designate rows and columns respectively.

The column sequences are actually particular cases of the generalized Fibonacci and Lucas sequences $\{F_{m,n}\}$, which satisfy the Fibonacci partial recurrence relation (Cook and Shannon 2006):

$$F_{m,n} = F_{m,n-1} + F_{m,n-2},\ m \geq 0,\ n > 2.$$

We now label the sequences of diagonal, row and partial column sums by $\{d_n\}$, $\{r_n\}$, $\{c_n\}$, respectively. We observe the sequences so generated in Table 5.4.2.

Table 5.4.1. A form of the Zeckendorf Triangle

1										
1	1									
2	1	2								
3	2	2	3							
5	3	4	3	5						
8	5	6	6	5	8					
13	8	10	9	10	8	13				
21	13	16	15	15	16	13	21			
34	21	26	24	25	24	26	21	34		
55	34	42	39	40	40	39	42	34	55	
89	55	68	63	65	64	65	63	68	55	89

Table 5.4.2. Sequences within the Zeckendorf Triangle

n	1	2	3	4	5	6	7	8	9	10	11
{dn}	1	1	3	4	9	13	25	38	68	106	182
{rn}	1	2	5	10	20	38	71	130	235	420	744
{cn}	1	2	6	15	40	104	273	714	1870	4895	12816
{bn}	1	1	4	9	25	64	169	441	1156	3025	7921

Table 5.4.3. Isosceles form of the Zeckendorf triangle

						1						
					1		1					
				2		1		2				
			3		2		2		3			
		5		3		4		3		5		
	8		5		6		6		5		8	
13		8		10		9		10		8		13

The $\{b_n\}$ sequence has been formed from the central column of the original isosceles form of the triangle in [2], as in Table 5.4.3.

Thus,

$$\{b_n\} \equiv \{z_{1,1}, z_{3,2}, z_{5,3}, z_{7,4}, z_{9.5}, \ldots\} \equiv \{1, 1, 4, 9, 25, 64, \ldots\} \qquad (5.4.1)$$

in which the $\{z_{i,j}\}$ are the elements of the isosceles form of the Zeckendorf triangle.

The $\{c_n\}$ sequence is formed from the cumulative partial sums of $\{b_n\}$:

$$\begin{aligned} b_n &= c_n - c_{n-1}, n > 1, \\ &= 2F_{2n-2} - b_{n-3}, \ n > 3 \\ &= F_n^2 . \end{aligned}$$

That is, by the repeated application of the first of these (with c_0 set to zero) we find that

$$\begin{aligned} c_n &= \sum_{j=1}^{n} F_n^2 \\ &= F_n F_{n+1} \end{aligned}$$

which can be confirmed from the table. We also observe too that, for $n \geq 2$,

$$d_{2n} = d_{2n-1} + d_{2n-2} ,$$

and

$$d_{2n+1} = d_{2n} + d_{2n-1} + F_{n+1}$$

Then, from the triangle, it can be seen that we get the recurrence relations

$$d_{2n-j} + d_{2n-j+1} + \delta_{1,j} F_n = d_{2n-j+2},\ j = 0,1, \tag{5.4.2}$$

in which $\delta_{i,j}$ is the Kronecker delta and $\{F_n\}$ are the Fibonacci numbers. Similarly, the $\{r_n\}$ is a Fibonacci convolution sequence (Hoggatt and Bicknell-Johnson 1977) where

$$5r_n = nF_{n+2} + (n+2)F_n \tag{5.4.3}$$

and

$$r_n + r_{n+1} + F_{n+2} = r_{n+2} \tag{5.4.4}$$

We note that (5.7.4) reduces to (5.4.3) with repeated use of the Fibonacci recurrence relation. The row numbers, $\{r_n\}$, were shown by Griffiths (2010) to be convolutions of the Fibonacci numbers. We can obtain another connection between the Fibonacci numbers and the Zeckendorf representations of the integers by defining another partial column sequence in Table 5.4.1, namely $\{k_{m,n}\}$, in which m identifies the row and n identifies the column as in Table 5.4.4.

Table 5.4.4. Partial column sums from Table 5.4.1

m→ n↓	1	2	3	4	5	6
1	1					
2	2	1				
3	4	2	2			
4	7	4	4	3		
5	12	7	8	6	5	
6	20	12	14	12	10	8

It can then be established that

$$k_{m,n} = k_{m-1,n-1} + k_{m-2,n-2}, \; m > 2, n > 2, \tag{5.4.5}$$

and within the columns

$$k_{m,n} = k_{m-1,n} + k_{m-1,} + F_n, \; n > m+1. \tag{5.4.6}$$

Many similar enumeration themes are unified in the Riordan Group (Shapiro et al 1991). Connections within the rows are left to the interested reader (Hilton and Pedersen 2012). Connections within the columns are related to the leading diagonals in Hoggatt's trimmed Pascal triangles (Hoggatt 1968).

5.5. Some Compositions Associated with Arbitrary Order Linear Recursive Sequences

A composition of the positive integer *n* is a vector $(a_1, a_2, ..., a_k)$ of which the components are the positive integers such that $a_1 + a_2 + ... + a_k = n$ (Heubach and Mansour 2009). If the vector has order *k*, then the composition is a *k*-part composition. In what follows $\gamma(n)$ will indicate summation over all the compositions $(a_1, a_2, ..., a_k)$ of *n*, the number of components being variable. The purpose of this section is to generalize to arbitrary order the second order Fibonacci and Lucas number results

$$n = \sum_{\gamma(n)} (-1)^{k-1} F_{2a_1} F_{2a_2} ... F_{2a_k} \tag{5.5.1}$$

$$L_{2n} = 2 + \sum_{\gamma(n)} (-1)^{k-1} \frac{n}{k} F_{2a_1} F_{2a_2} \ldots F_{2a_k} \qquad (5.5.2)$$

and similar expressions Ferns 1969; Hilton 1974; Hoggatt and Lind 1968, 1969; Horadam 1965; Moser and Whitney 1961; Zeitlin 1965). Ideas for further extensions may be found in Sarvate and Zhung 2014).

To achieve our purpose we consider linear recursive sequences of arbitrary order r, $\{w_{s,n}^{(r)}\}$, which satisfy the homogenous recurrence relation

$$w_{s,n+r}^{(r)} = \sum_{j=1}^{r} (-1)^{j+1} P_{r,j} w_{s,n+r-j}^{(r)} \qquad (5.5.3)$$

for $s = 0,1,\ldots,r\text{-}1$ and $n > 1$, with suitable initial values for $w_{s,n}^{(r)}$, $n = 0,1,2,\ldots,r\text{-}1$. Modifying Williams (1972) we let $\alpha_{r,j}$ be the r roots, assumed distinct, of the associated auxiliary equation,

$$x^r = \sum_{j=1}^{r} (-1)^{j+1} P_{r,j} x^{r-j} \qquad (5.5.4)$$

where the roots are related to roots of unity:

$$\alpha_{r,j}^{n} = \frac{1}{r} \sum_{k=0}^{r-1} w_{k,r+n}^{(r)} d^k \zeta_r^{-jk} \qquad (5.5.5)$$

for suitable real d, and in which

$$\zeta_r = \exp\left(\frac{2\pi i}{r}\right).$$

When $r = 2$ and $n = 1$, Equation (5.5.5) becomes

$$\alpha_{2,j} = \frac{1}{2}\left(w_{0,3}^{(2)} + (-1)^j dw_{1,3}^{(2)}\right)$$

which yields

$$w_{0,n+2}^{(2)} = L_n \text{ and } w_{1,n+2}^{(2)} = F_n.$$

We shall also have occasion to use

$$\sum_{j=1}^{r} \varsigma_r^{-tj} = r\delta_{t0}$$

in which δ_{tj} is the Kronecker delta.

Proof:

$$\sum_{j=1}^{r} \varsigma_r^{-tj} = \begin{cases} r & \textit{if } t = 0, \\ \varsigma_r^{-t} \dfrac{\left(1 - \varsigma_r^{-tr}\right)}{\left(1 - \varsigma_r^{-t}\right)} & \textit{otherwise}, \end{cases}$$
$$= \begin{cases} r & \textit{if } t - 0, \\ 0 & \textit{otherwise} \end{cases}$$
$$= r\delta t_{to}$$

Since

$$\varsigma_r^{-tr} = \exp\left(\frac{-2\pi ir}{r}\right)t$$
$$= \exp(-2\pi i)t$$
$$= 1.$$

The next result we need is an obvious extension of the so-called Binet general formulas for Fibonacci and Lucas numbers:

$$w_{s,r+n}^{(r)} = d^{-s} \sum_{j=1}^{r} \alpha_{r,j}^{n} \varsigma_r^{sj},\ s = 0,1,\ldots,r-1. \qquad (5.5.6)$$

***Proof*:**

$$\sum_{j=1}^{r} \alpha_{r,j}^{n} \varsigma_r^{sj} = \frac{1}{r} \sum_{k=0}^{r-1} w_{k,r+n}^{(r)} d^k \sum_{j=1}^{r} \varsigma_r^{(t-k)j}$$
$$= \frac{1}{r} w_{t,r+n}^{(r)} d^t r,$$

from which the result follows.

Examples of (5.3.5) are the Binet expressions for the general terms

$$w_{0,n+2}^{(2)} = \sum_{j=1}^{2} \alpha_{2,j}^{n}$$

for the Lucas numbers, and

$$w_{1,n+2}^{(2)} = d^{-1} \sum_{j=1}^{2} \alpha_{2,j}^{n}$$

for the Fibonacci numbers. This is another way of saying that the $w_{s,n}^{(r)}$ of Equation (5.5.6) satisfy the recurrence relation in (5.5.3); that is,

$$\sum_{k-1}^{r} P_{r,k}\left(d^{-s} \sum_{m=1}^{r} (-1)^{k+1} \alpha_{r,j}^{n-k} \varsigma_r^{sm} \right) = d^{-s} \sum_{m=1}^{r} \left(\sum_{k=1}^{r} (-1)^{k+1} \alpha_{r,m}^{r-k} P_{r,k} \right) \alpha_{r,m}^{n-r} \varsigma_r^{sm}$$
$$= d^{-s} \sum_{m=1}^{r} \left(\alpha_{r,m}^{r}\right) \alpha_{r,m}^{n-r} \varsigma_r^{s,m}$$
$$= w_{s,n+r}^{(r)}$$

as required.

In order to generalize the result (5.5.1) it is first necessary to generalize the result

$$F_{2n} = 3F_{2n-2} - F_{2n-4} \tag{5.5.7}$$

on which it depends [4].

$$w_{s,rn}^{(r)} = \sum_{j=1}^{r} s(r,r,j) w_{s,r(n-j)}^{(r0} \tag{5.5.8}$$

in which the $s(r,m,j)$ are the symmetric functions of the $\alpha_{r,i}^{m}$, $i = 1,2,\ldots,r$, taken j at a time (MacMahon 1915).

$$s(r,m,j) = \sum \alpha_{r,i_1}^{m} \alpha_{r,i_2}^{m} \ldots \alpha_{r,i_j}^{m}$$

in which the sum is over a distinct cycle of $\alpha_{r,i}^{m}$, we set $s(r,m,0) = 1$. For instance,

$$s(3,m,1) = \alpha_{3,1}^{m} + \alpha_{3,2}^{m} + \alpha_{3,2}^{m}$$

$$s(3,m,2) = (\alpha_{3,1}\alpha_{3,2})^{m} + (\alpha_{3,2}\alpha_{3,3})^{m} + (\alpha_{3,3}\alpha_{3,1})^{m}$$

$$s(3,m,3) = (\alpha_{3,1}\alpha_{3,2}\alpha_{3,3})^{m}.$$

Proof of (4.2):
For notational convenience we write (5.5.6) in the form

$$w_{s,n}^{(r)} = \sum_{j=1}^{r} A_{s,j} \alpha_{r,j}^{n}$$

in which the $A_{s,j}$ are determined by the initial values of $w_{s,n}^{(r)}$.

Now

$$\sum_{j=1}^{r}(-1)^{j+1}s(r,r,j)w_{s,r(n-j)}^{(r)} = \sum_{j=1}^{r}(-1)^{j+1}s(r,r,j)\sum_{i=1}^{r}A_{s,i}\alpha_{r,i}^{rn-rj}$$

$$= \sum_{j=1}^{r}\alpha_{r,j}^{r}\sum_{i=1}^{r}A_{s,i}\alpha_{r,i}^{rn-r} - \sum_{\substack{j,k=1\\ j\neq k}}^{r}\alpha_{r,j}^{r}\alpha_{r,k}^{r}\sum_{i=1}^{r}A_{s,i}\alpha_{r,i}^{rn-2r} + \ldots + (-1)^{r+1}\left(\alpha_{r,1}^{r}\alpha_{r,2}^{r}\ldots\alpha_{r,r}^{r}\right)\sum_{i=1}^{r}A_{s,i}\alpha_{r,i}^{rn=rr}$$

$$= \sum_{j=1}^{r}A_{s,j}\alpha_{r,j}^{rn} + \sum_{\substack{j,k=1\\ j\neq k}}^{r}A_{s,j}\alpha_{r,j}^{rn-r}\alpha_{r,k}^{r} - \sum_{\substack{j,k=1\\ j\neq k}}^{r}A_{s,j}\alpha_{r,j}^{rn-r}\alpha_{r,k}^{r} - \sum_{\substack{i,j,k=1\\ i\neq j\neq k}}^{r}A_{s,i}\alpha_{r,i}^{rn-2r}\alpha_{r,j}^{r}\alpha_{r,k}^{r} + \ldots$$

$$= \sum_{j=1}^{r}A_{s,j}\alpha_{r,j}^{rn}$$

$$= w_{s,rn}^{(r)}.$$

For example,

$$w_{1,n}^{(2)} = s(2,2,1)w_{s,2n-2}^{(2)} - s(2,2,2)w_{2,2n-4}^{(2)}$$

$$= 3w_{s,2n-2}^{(2)} - w_{2,2n-4}^{(2)}$$

as in (5.5.7) above.

We next define another sequence $\left\{u_{s,n}^{(r)}\right\}$ by

$$u_{s,n}^{(r)} = \sum_{\gamma(n)}(-1)^{k-1}w_{s,a_1}^{(r)}w_{s,a_2}^{(r)}\ldots w_{s,a_k}^{(r)} \tag{5.5.9}$$

and show that the formal power series generating functions

$$U(x)=\sum_{n=1}^{\infty}u_{s,n}^{(r)}x^n$$

and

$$W(x)=\sum_{n=1}^{\infty}w_{s,n}^{(r)}x^n$$

are related by

$$U(x)=\frac{W(x)}{1+W(x)}. \tag{5.5.10}$$

Proof:
The coefficient of x^n in

$$\left(W(x)\right)^k=\left(w_{s,1}^{(r)}x+w_{s,2}^{(r)}x^2+...\right)^k$$

is the sum of terms $w_{s,a_1}^{(r)}, w_{s,a_2}^{(r)},...$ in which $a_1+a_2+...=n$, and the number of summands in this last sum is *k*. Thus

$$\begin{aligned}U(x)&=\sum_{n=1}^{\infty}u_{s,n}^{(r)}x^n\\&=\sum_{n=1}^{\infty}\sum_{\gamma(n)}(-1)^{k-1}w_{s,a_1}^{(r)}...w_{s,a_k}^{(r)}x^n\\&=\sum_{k=1}^{\infty}-\left(-W(x)\right)^k\\&=\frac{W(x)}{1+W(x)}.\end{aligned}$$

Lemma 1: If

$$h(x) = f(x)W(x),$$

then

$$h(x) = \big(f(x) + h(x)\big)U(x).$$

Proof:

$$U(x) = \frac{W(x)}{1 + W(x)}$$
$$= \frac{f(x)W(x)}{f(x) + f(x)W(x)}$$
$$= \frac{h(x)}{f(x) + h(x)}.$$

Lemma 2: If

$$f(x) = \sum_{j=0}^{r} (-1)^{r-j} P_{r,j} x^j, \quad P_{r,0} = 1,$$

and

$$h(x) = \sum_{j=0}^{r} (-1)^{r-j} h_j x^j, \quad h_0 = 0,$$

then

$$h_j = \sum_{m=1}^{j} (-1)^m P_{r,j-m} w_{s,m}^{(r)}.$$

Proof:

$$
\begin{aligned}
h(x) &= W(x)f(x) \\
&= \sum_{n=1}^{\infty} w_{s,n}^{(r)} x^n \sum_{j=0}^{r} (-1)^{r-j} P_{r,j} x^j \\
&= \sum_{j=1}^{r} \left(\sum_{m=1}^{j} (-1)^{r-j+m} P_{r,j-m} w_{s,m}^{(r)} \right) x^j + \sum_{j=1}^{\infty} \left(\sum_{m=0}^{r} (-1)^m P_{r.r-m} w_{s,j+m}^{(r)} \right) x^{r+j} \\
&= \sum_{j=1}^{r} (-1)^{r-j} \left(\sum_{m=1}^{j} (-1)^m P_{r,j-m} w_{r,m}^{(r)} \right) x^j
\end{aligned}
$$

since for $P_{r,0} = 1$,

$$
0 = \sum_{m=0}^{r} (-1)^m P_{r,r-m} w_{s,j+m}^{(r)} .
$$

Thus

$$
f(x) + h(x) = \sum_{j=0}^{r} (-1)^{r-j} \left(P_{r,j} + h_j \right) x^j
$$

and since

$$
h(x) = \left(f(x) + h(x) \right) U(x)
$$

we have that $u_{s,n}^{(r)}$ satisfies the linear recurrence relation

$$
u_{s,n}^{(r)} = \sum_{j=1}^{r} (-1)^{j+1} \left(P_{r,j} + h_j \right) u_{s,n-j}^{(r)} .
$$

We have proved that if

$$w_{s,n}^{(r)} = \sum_{j=1}^{(r)} (-1)^{j+1} P_{r,j} w_{s,n-j}^{(r)},$$

and

$$u_{s,n}^{(r)} = \sum_{\gamma(n)} (-1)^{k-1} w_{s,a_1}^{(r)} \ldots w_{s,a_k}^{(r)},$$

then

$$u_{s,n}^{(r)} = \sum_{j=1}^{(r)} (-1)^{j+1} \left(P_{r,j} + h_j\right) w_{s,n-j}^{(r)},$$

in which

$$h_j = \sum_{m=1}^{j} (-1)^m P_{r,j-m} w_{s,m}^{(r)}$$

and $\gamma(n)$ indicates summation over all the compositions $(a_1, a_2, \ldots, a_k)$ of n.

The corollary to this is that if

$$w_{s,rn}^{(r)} = \sum_{j=1}^{(r)} (-1)^{j+1} s(r,r,j) w_{s,r(n-j)}^{(r)}$$

and

$$u_{s,n}^{(r)} = \sum_{\gamma(n)} (-1)^{k-1} w_{s,ra_1}^{(r)} \ldots w_{s,a_k}^{(r)},$$

then

$$u_{s,n}^{(r)} = \sum_{j=1}^{(r)} (-1)^{j+1} \left(s(r,r,j) + k_j \right) u_{s,n-j}$$

in which

$$k_j = \sum_{m=1}^{j} (-1)^m s(r, rj - m) w_{s,rm}^{(r)}.$$

That the corollary generalizes the result in (5.5.1) can be seen if we let $r = 2$ and

$$u_{1,n}^{(2)} = \sum_{\gamma(n)} (-1)^{k-1} w_{1,a_1}^{(2)} \ldots w_{1,a_k}^{(2)}$$

and

$$\begin{aligned}
u_{1,n}^{(2)} &= \sum_{j=1}^{2} (-1)^{j+1} \left(s(2,2,j) + k_j \right) u_{1,n-j}^{(2)} \\
&= \left(s(2,2,1) + k_1) \right) u_{1,n-1}^{(2)} - \left(s(2,2,2) + k_2 \right) u_{1,n-2}^{(2)} \\
&= \left(s(2,2,1) - s(2,2,0) F_2 \right) u_{1,n-1}^{(2)} - \left(s(2,2,2) - s(2,2,1) F_2 + s(2,2,0) F_4 \right) u_{1,n-2}^{(2)} \\
&= (3-1) u_{1,n-1}^{(2)} - (1 - 3 + 3) u_{1,n-2}^{(2)} \\
&= 2 u_{1,n-1}^{(2)} - u_{1,n-2}^{(2)} \\
&= n.
\end{aligned}$$

We now note that result (5.5.1) can be considered as a particular part of some theory discussed by Jerbic (1968) whose (corrected) result is that

$$y_n = n^k$$

satisfies

$$0 = \sum_{j=0}^{k+1} (-1) \binom{k+1}{j} y_{k+n+1-j}. \tag{5.5.11}$$

For example,

$$y_n = n^0$$

satisfies

$$0 = y_{n+1} - y_n,$$

and

$$y_n = n^1$$

satisfies

$$0 = y_{n+2} - 2y_{n+1} + y_n,$$

and

$$y_n = n^2$$

satisfies

$$0 = y_{n+3} - 3y_{n+2} + 3y_{n+1} - y_n.$$

We then get a generalization of the corollary in the previous section; namely,

$$w_{s,n}^{(r)} = n^{r-1} \tag{5.5.12}$$

is a solution of the difference equation

$$w_{s,n+r}^{(r)} = \sum_{j=1}^{r} (-1)^{j+1} P_{r,j} w_{s,n+r-j}^{(r)} \tag{5.5.13}$$

when

$$P_{r,j} = \binom{r}{j}. \tag{5.5.14}$$

In order to prove (5.5.14) we need the result

$$0 = \sum_{j=0}^{r} (-1)^j \left(\binom{r}{j} (n+r-j)^{r-1} \right). \tag{5.5.15}$$

Proof:
We use induction on r:

$$r = 1: \quad (n+1)^0 - n^0 = 0;$$

$$r - 1: \quad (n+2) - (2(n+1) + n = 0$$

$$r = 3: \quad (n+3)^2 - 3(n+2)^2 + 3(n+1)_2 - n^2 = 0.$$

Assume the result is true for $r = 4,5,\ldots,k$-1. Then

$$\begin{aligned}\sum_{j=0}^{k}(-1)^{j}\binom{k}{j}(n+k-j)^{k-1} &= \sum_{j=0}^{k}(-1)^{k-j}\binom{k}{j}(n+j)^{k-1}\\ &= \sum_{j=0}^{k-1}(-1)^{k-j}\binom{k-1}{j}(n+j)^{k-1}+\sum_{j=0}^{k}(-1)^{k-j}\binom{k-1}{j-1}(n+j)^{k-1}\\ &= \sum_{j=0}^{k-1}(-1)^{k-j}\binom{k-1}{j}(n+j)^{k-2}(n+j)-\sum_{j=0}^{k}(-1)^{k-j}\binom{k-1}{j-1}(n+j)^{k-2}(n+j+1)\\ &= \sum_{j=0}^{k-1}(-1)^{k-j}\binom{k-1}{j}(n+j)^{k-2}j-\sum_{j=0}^{k}(-1)^{k-j}\binom{k-1}{j}(n+j+1)^{k-2}j\\ &= 0\end{aligned}$$

in which we have used the inductive hypothesis:

$$\begin{aligned}\sum_{j=0}^{k-1}(-1)^{k-j}\binom{k-1}{j}(n+j)^{k-2}j &= \sum_{j=1}^{k}(-1)^{k-j}(k-j)\binom{k-1}{j-1}(n+j)^{k-2}\\ &= -\sum_{j=0}^{k-1}(-1)^{k-j}(k-j-1)\binom{k-1}{j}(n+j+1)^{k-2}\\ &= \sum_{j=0}^{k-1}(-1)^{k-j}\binom{k-1}{j}(n+j+1)^{k-2}j\end{aligned}$$

This completes the proof of (5.5.15). It is not likely that the lemma is new since "any result about the binomial coefficients is in the literature" (Krall 1960).

The proof of (5.5.12) now follows if we replace $(n + r - j)$ in (6.3) by $\left(w_{n+r-j}^{(r)}\right)^{1/(r-1)}$. An example of this was seen previously, namely, $u_{1,n}^{(2)} = 2u_{1,n-1}^{(2)} - u_{1,n-2}^{(2)} = n.$

Finally in this section, in order to generalize (5.5.2), we let

$$w_{s,n}^{(r)} = \sum_{\gamma(n)} \frac{(-1)^{k-1}}{k} u_{s,a_1}^{(r)} \ldots u_{s,a_k}^{(r)}. \tag{5.5.16}$$

Then

$$\sum_{n=1}^{\infty} w_{s,n}^{(r)} x^n = \sum_{n=1}^{\infty} \sum_{\gamma(n)} \frac{(-1)^{k-1}}{k} u_{s,a_1}^{(r)} \ldots u_{s,a_k}^{(r)} x^n$$

$$= \sum_{k=1}^{\infty} - \left(- \sum_{n=1}^{\infty} u_{r,n}^{(r)} x^n \right) / k$$

$$= \ln \left(1 + \sum_{n=1}^{\infty} u_{s,n}^{(r)} x^n \right)$$

$$= \ln \left(\sum_{n=0}^{\infty} u_{s.n}^{(r)} x^n \right)$$

or

$$\sum_{n=0}^{\infty} u_{s,n}^{(r)} x^n = \exp \left(\sum_{n=1}^{\infty} w_{s,n}^{(r)} x^n \right) \tag{5.5.17}$$

which is satisfied by

$$u_{s,n}^{(r)} = \frac{w_{1,n+2}^{(r)}}{n}.$$

Thus

$$w_{1,n+2}^{(r)} = \sum_{\gamma(n)} (-1)^{k-1} \frac{n}{k} w_{0,a_1}^{(r)} \ldots w_{0,a_k}^{(r)}$$

as a generalization of (5.5.2).

Gratitude is expressed to Professor Henry W Gould of West Virginia University for noting the Krall reference to us some years ago.

Chapter 6

CONCLUSION

This chapter brings together some of the preceding ideas in the setting of generalized integers and arbitrary order recursive sequences which shed light on the nature of the relationship between the Fibonacci and Lucas sequences in the second order case.

In turn, they suggest further generalizations either algebraically or computationally particularly with the aid of the invaluable *Encyclopedia of Integer Sequences* which should be part of the armory of all number theorists.

This leads to some final comments on the role of number theory and its close relatives in combinatorics, geometry and statistics as a modern *quadrivium* for 21st century liberal education in undergraduate education. For those who want to explore some of these mathematical ideas further in Latin (via Italian) or Greek see Baldassarre Boncompagni-Ludovisi (1854) or Verner Hoggatt - Μεταφραση: Ανδρεασ Ν. Ριλιππου & Γιωργοσ Ν. Ριλιππου (1983).

As well as drawing together the various strands expounded in the book, several topics for further investigation are outlined. These are suggested specifically in some sections but others arise from the new knowledge which arises from critical reflections on existing results.

Comments on the place of mathematics in general for a citizen of the 21st Century, let alone for the fun in the puzzles that mathematics generates (Spencer 2017), number theory in particular, in the liberal arts as a foundation for 21st century education are pertinent to much of the ongoing discussion in the media as well as within professional societies on generic skills and sustained employability. In some ways this book has aims similar to "reverse mathematics" (Stillwell 2018) whose recent book aims to "engage advanced undergraduates and all mathematicians interested in the foundations of mathematics" by "using a minimum of mathematical logic in a well-motivated way".

Thus, Larcombe (2017) refers to Leonard Kronecker's description of mathematicians as "poets in truth", and Paul Halmos (1968): "Talk to a painter (I did) and talk to a mathematician and you'll be amazed at how similarly they react. Almost every aspect of the life and of the art of a mathematician has its counterpart in painting, and vice versa".

6.1. Related Topics

Further related investigations can include

- establishing a general form for $F_n(a)$ as a polynomial in a as in Table 4.3.2;
- determining intersections of the generalized sequences in Table 4.3.4;
- exploring identities analogous to those of the ordinary Fibonacci and Lucas numbers, including the use of fuzzy logic (Atanassov 2017; Atanassov et al 1997);
- using simple factorials to calculate generalized Pascal–Fibonacci numbers and Pascal-type triangles;
- extending the ideas to spirals (Zahn 2017) and cylinders (Atanassov and Shannon 2008) ;

- finding the missing patterns inherent in Table 4.3.4 where some of the sequences do not appear in the *Online Encyclopedia of Integer Sequences*;
- developing generating functions and polynomials for these generalizations, particularly analogs of Fibonomials (Gould 1969) and Lucasnomials (Ballot 2017);
- developing $w_{m,n} = w_{m-1,n} + w_{m,n-1}$ of Section 5.3;
- generalizations of the Pascal–Fibonacci numbers which can be made with suitable generalizations of Pascal's triangle (Wong and Maddocks 1975);
- extending Horadam's notation to higher order reurrences:

Table 6.1.1. Integer and polynomial sequences $\{w_{2,n}(a,b;p,q)\}$

a	b	p	q	wn	Sequence
0	1	1	-1	Fn	Fibonacci
2	1	1	-1	Ln	Lucas
0	1	2	-1	Pn	Pell
1	3	2	-1	Qn	Pell-Lucas
0	1	x+2	+1	Bn(x)	Morgan-Voyce Even Fibonacci
1	1	x+2	+1	bn(x)	Morgan-Voyce Odd Fibonacci
2	x+2	x+2	+1	Cn(x)	Morgan-Voyce Even Lucas
-1	1	x+2	+1	cn(x)	Morgan-Voyce Odd Lucas
0	1	1	-x	Jn(x)	Jacobsthal-Fibonacci
2	1	1	-x	jn(x)	Jacobsthal-Lucas
0	1	x	+1	Vn(x)	Vieta-Fibonacci
2	x	x	+1	vn(x)	Vieta-Lucas
a	a+d	2	+1	An	Arithmetic
a	ar	r	0	Gn	Geometric

Table 6.1.1 illustrates the efficiency and effectiveness of Horadam's notation in placing these sequences in perspective for comparisons. More recent further generalizations relevant to the subsequent development in this book include coupled and multiplicative Fibonacci sequences (Singh

2012; Atanassov et al. 2015) and particularly the table in the second scheme (Singh 2010; Shannon and Pruitt, 2017).

This section and the next section of this final chapter also mention two extensions for the interested reader to pursue, namely,

- generalized integers and arbitrary order recursive sequences, and
- aspects of Horadam's generalized sequence of numbers, $\{w_n(a,b;p,q)\}$, defined by the second order linear homogeneous recurrence relation

$$w_n = pw_{n-1} - qw_{n-2}, \; n > 2.$$

with initial conditions $w_0 = a, w_1 = b$ (Horadam 1965a) with fundamental properties (Horadam 1965b, 1967, 1968). Carlitz (1954) had also previously looked at some of the arithmetic properties of generalized sequences and functional difference equations (Carlitz 1964a), and we build on some of his ideas here too.

We now outline Fermatian numbers as examples of generalized integers and consider their multinomial coefficients as analogues of ordinary arithmetical functions. Each example is related to a Sloane sequence. We here consider Fermatian numbers as examples of generalized integers and their multinomial coefficients as analogues of ordinary arithmetical functions. This chapter is partly expository and partly indicating new problems to extend some of the earlier ideas in this book.

Other approaches, such as Daykin (1963) and Mollie Horadam (1971), were to start with generalized primes as a foundation for generalized integers. Here though the approach is to start with divisibility sequences, the elements of which we consider as generalized integers, and to call some of their elements as generalized primes. That is, u_p is a generalized prime (modulo a divisibility sequence) if its only divisors are itself and unity (within the system).

Our so-called generalized integers cannot necessarily be represented as a product of distinct generalized primes, as we shall see. Thus new analogues of the classical arithmetical functions are needed in order to study these divisibility sequences as generalizations of integers.

Other approaches of Cohen (1975), Brent and Cohen (1989) and McDaniel (1974) have been to focus on Gaussian integers as generalized integers in the context of analogues of number theory problems. We start with the sequence of Fermatian numbers $\{\underline{q}_n\}$ which may be defined in terms of real numbers q such that $\underline{q}_n$ is the nth Fermatian number of index q (cf. Sylvester 1912; Childs 1979)

$$\underline{q}_n = \begin{cases} -q^n \underline{q}_{-n} & (n < 0) \\ 1 + q + q^2 + \ldots q^{n-1} & (n > 0) \\ 1 & (n = 0) \end{cases} \tag{6.1.1}$$

so that

$$\underline{1}_n = n. \tag{6.1.2}$$

and

$$\underline{1}_n! = n!, \tag{6.1.3}$$

where

$$\underline{q}_n! = \underline{q}_n \underline{q}_{n-1} \ldots \underline{q}_1 . \tag{6.1.4}$$

For example, if we consider the Fermatian numbers of index 2, we have $\underline{2}_2 = 3$, $\underline{2}_3 = 7$, $\underline{2}_4 = 15$, $\underline{2}_6 = 63$, $\underline{2}_8 = 255$, so that $\underline{2}_2$, $\underline{2}_3$ and $\underline{2}_4$ are generalized Fermatian primes, and $\underline{2}_6 = (\underline{2}_2)^2\underline{2}_3$, but $\underline{2}_8$ cannot be represented as a product of Fermatian numbers of index 2. Some properties of these numbers may be found in Shannon (2004) and Carlitz and Moser (1966).

Table 6.1.2. First 10 Fermatian numbers of the first 10 indices

Index q↓	1	2	3	4	5	6	7	8	9	10
1	1	2	3	4	5	6	7	8	9	10
2	1	3	7	15	31	63	127	255	511	1023
3	1	4	13	40	121	364	1093	3280	9841	29524
4	1	5	21	85	341	1365	5461	21845	87381	349525
5	1	6	31	156	781	3906	19531	97656	488281	2441406
6	1	7	43	259	1555	9331	55987	335923	2015539	12093235
7	1	8	57	400	2801	19608	137257	960800	6725601	47079208
8	1	9	73	585	4681	37449	299593	2396745	19173961	153391689
9	1	10	91	820	7381	66430	597871	5380840	48427561	435848050
10	1	11	111	1111	11111	111111	1111111	11111111	111111111	1111111111

Carlitz (1948) has also used $\underline{q}_n$ in the development of q-Bernoulli numbers and polynomials but he adopted different notation, namely [x], but as [x] used to be used commonly for the greatest integer function and as Carlitz (1940) himself used [k] to mean other things, it is felt that $\underline{z}_n$ is less confusing.

The first ten Fermatian numbers of the first ten indices are displayed in Table 6.1.2.

It is obvious from looking along the rows that the numbers satisfy the first order inhomogenous recurrence relation, well known for the Repunits when $q = 10$,

$$\underline{q}_n = q\underline{q}_{n-1} + 1 \qquad (6.1.5)$$

and the second order homogeneous recurrence relation

$$\underline{q}_n = (q+1)\underline{q}_{n-1} - q\underline{q}_{n-2}. \qquad (6.1.6)$$

Of more interest are the relations among the row, columns and various diagonals to be found with the connections to Sloane (1995) as indicated in the last column and row of Table 6.1.3.

Table 6.1.3. Connections with Sloane integer sequences

q	3	4	5	6	7	8	9	10	Sloane
1	3	4	5	6	7	8	9	10	A000027
2	7	15	31	63	127	255	511	1023	A000225
3	13	40	121	364	1093	3280	9841	29524	A003462
4	21	85	341	1365	5461	21845	87381	349525	A002450
5	31	156	781	3906	19531	97656	488281	2441406	A003463
6	43	259	1555	9331	55987	335923	2015539	12093235	A003464
7	57	400	2801	19608	137257	960800	6725601	47079208	A023000
8	73	585	4681	37449	299593	2396745	19173961	153391689	A023001
9	91	820	7381	66430	597871	5380840	48427561	435848050	A002452
10	111	1111	11111	111111	1111111	11111111	111111111	1111111111	A002275
Sloane	**002061**	**053698**	**053699**	**053700**	**053716**	**053717**	**102909**	**103623**	

If

$$T_n = \sum_{k=1}^{r} T_{n-k},$$

then T_n is the sum of the rising diagonals of the multinomial triangle generated by $\underline{z}_n^r$ (Cox et al 1960).

Hoggatt and Bicknell (1969) proved that, for the general *r*-nomial triangle induced by the expansion $\underline{q}_n^r$ (n = 0,1,2,3,...), by letting the *r*-nomial triangle be left-justified and by taking sums from the left edge and jumping up p and over 1 entry until out of the triangle that

$$T_n = \sum_{k=0}^{\left\lfloor \frac{1}{2} n(r-1) \right\rfloor} \left\{ {n-k \atop k} \right\}_r ,$$

where

$$\underline{z}_r^n = \sum_{j=0}^{n(r-1)} \begin{Bmatrix} n \\ j \end{Bmatrix}_r x^j,$$

and the r-nomial coefficient $\begin{Bmatrix} n \\ j \end{Bmatrix}_r$ is the entry in the n-th row and j-th column of the generalized Pascal triangle (Bondarenko 1993). Thus,

$$\left(1 - x\underline{\left(x^p\right)}_r\right)^{-1} = \sum_{n=0}^{\infty} \left(x\underline{\left(x^p\right)}_r\right)^n$$
$$= \sum_{n=0}^{\infty} \left(\sum_{k=0}^{\left\lfloor \frac{n(r-1)}{p+1} \right\rfloor} \begin{Bmatrix} n-kp \\ k \end{Bmatrix}_r \right) x^n$$

which, when $p = 1$, is a generating function for $\{T_n\}$ with suitable initial values. Here,

$$\underline{\left(x^p\right)}_r = 1 + x^p + x^{2p} + \ldots + x^{p(r-1)}$$

so that the notation is quite versatile. The corresponding row and column sequences, $\{\underline{z}_n\}_{n=1}^{\infty}$, $\{\underline{z}_n\}_{z=1}^{\infty}$, are obvious from their construction, but the sequence,

$$\left\{ \sum_{n=1}^{z-1} \underline{(z-n)}_n \right\} \equiv \{1,3,6,11,21,45,105,315,1058,\ldots\},$$

formed from adding along the forward diagonals, does not seem to be well-known. More generally, following Gould (1969) we define Fontené-Ward multinomial coefficients by

$$\begin{Bmatrix} n \\ s_1, s_2, \ldots, s_k \end{Bmatrix} = \frac{u_n!}{u!_{s_1} u_{s_2}! \ldots u_{s_k}!} \tag{6.1.7}$$

in which $\{u_n!\}$ is an arbitrary sequence of real or complex numbers such that $u_n \neq 0, n \geq 1, u_0 = 0, u_1 = 1,$

$$u_n! = u_n u_{n-1} \ldots u_1 \tag{6.1.8}$$

which is satisfied by the Fibonacci sequence (Hoggatt 1967) and by the Fermatian numbers $\{\underline{q}_n\}$ which can in turn be related to fermionic *p*-adic numbers with their rich properties (Dolgy et al. 2016). For the Fermatian numbers the binomial coefficient can be written as

$$\begin{Bmatrix} n \\ s_1, s_2 \end{Bmatrix}_q = \frac{\underline{q}_n!}{\underline{q}_{s_1}! \underline{q}_{s_2}!} \tag{6.1.9}$$

or more simply

$$\begin{Bmatrix} n \\ s \end{Bmatrix} = \frac{\underline{q}_n!}{\underline{q}_s! \underline{q}_{n-s}!}$$

$$= \frac{\underline{q}_n \underline{q}_{n-1} \cdots \underline{q}_{n-s+1}}{\underline{q}_s!}$$

$$= \frac{\underline{q}_n^{\bar{s}}}{\underline{q}_s!}$$

in which we have the Fermatian equivalent of the rising factorial coefficient defined by

$$\underline{q}_n^{\bar{s}} = \underline{q}_n \underline{q}_{n-1} \ldots \underline{q}_{n-s+1}.$$

The recurrence relation for ordinary multinomial coefficients can be expressed as (Hoggatt and Alexanderson 1971)

$$\binom{n}{s_1,\ldots,s_r} = \sum_{j=1}^{r} \binom{n-1}{s_1\delta_{1,j},\ldots,s_r\delta_{r,j}},\ n = \sum_{i=1}^{r} s_i,$$

in which $\delta_{i,j}$ represents the Kronecker delta. In order to determine the Fermatian equivalent we need some preliminary results for arbitrary order recursive sequences. We denote a square matrix of order r by

$$M_n^{(r)} = \begin{bmatrix} U_{1,n+1}^{(r)} & U_{2,n+1}^{(r)} & \cdots & U_{r,n+1}^{(r)} \\ U_{1,n+2}^{(r)} & U_{2,n+2}^{(r)} & \cdots & U_{r,n+2}^{(r)} \\ & & \cdots & \\ U_{1,n+r}^{(r)} & U_{2,n+r}^{(r)} & \cdots & U_{r,n+r}^{(r)} \end{bmatrix} \quad (6.1.10)$$

in which we have r basic sequences of arbitrary order r

$$U_{s,n}^{(r)} = \sum_{j=1}^{r} (-1)^{j+1} P_{r,j} U_{s,n-j}^{(r)},\ n > r, \quad (6.1.11)$$

and initial terms

$$U_{s,n}^{(r)} = \delta_{s,n},\ n = 1,2,\ldots,r,$$

with arbitrary integers $P_{r,j}$ (Shannon 2013; Philippou 1983). It can then be established (Shannon 1974) that

$$M_n^{(r)} M_m^{(r)} = M_{m+n}^{(r)}$$

from which we obtain on equating the last row and last column that

$$U^{(r)}_{r,r+n} = \sum_{j=0}^{r-1} U^{(r)}_{r-j,r+m} U^{(r)}_{r,r+n-m-j}.$$

Then

$$\begin{aligned}\sum_{j=1}^{r} \left\{ \begin{matrix} n-1 \\ s_1\delta_{1,j},\dots,s_r\delta_{r,j} \end{matrix} \right\} U^{(r)}_{r-j+1,r+m} &= \frac{\underline{q}_{n-1}!\left(U^{(r)}_{r,n-m}U^{(r)}_{r,r+m} + \dots + U^{(r)}_{r,n-m-r+1}U^{(r)}_{1,r+m}\right)}{\underline{q}_{s_1}!\dots\underline{q}_{s_r}!} \\ &= \frac{\underline{q}_{n-1}!U^{(r)}_{r,n}}{\underline{q}_{s_1}!\dots\underline{q}_{s_r}!} \\ &= \left\{ \begin{matrix} n-1 \\ s_1,\dots,s_r \end{matrix} \right\} \end{aligned} \qquad (6.1.12)$$

with

$$\begin{aligned}\sum_{i=1}^{r} s_i &= \sum_{i=1}^{r} (n-m-i+1) \\ &= r(n-m+1) - \tfrac{1}{2}r(r+1) \\ &= n.\end{aligned}$$

For example, when $r = 2$, $n = 2k + 1$, $m = k$, $s_1 = k+1$, $s_2 = k$, then if we replace the Fermatian numbers by Fibonacci numbers, (6.1.9) becomes in our notation

$$\begin{aligned}\left\{ \begin{matrix} 2k+1 \\ k \end{matrix} \right\} &= U^{(2)}_{2,k+2} \left\{ \begin{matrix} 2k \\ k \end{matrix} \right\} + U^{(2)}_{1,k+2} \left\{ \begin{matrix} 2k \\ k-1 \end{matrix} \right\} \\ &= U^{(2)}_{2,k+2} \left\{ \begin{matrix} 2k \\ k \end{matrix} \right\} - P_{2,2} U^{(2)}_{2,k+1} \left\{ \begin{matrix} 2k \\ k-1 \end{matrix} \right\}\end{aligned}$$

which agree with corresponding results in Hoggatt (1967) and Alwyn Horadam (1965). When the Fermatian numbers are replaced by ordinary integers and $P_{2,1} = 1$, $P_{2,0} = 0$ in (6.1.11), we get the ordinary binomial partial recurrence relation.

There is still much discovery that can be done with generalized integers apart from studying connections arising from considerations of the diagonals in Table 6.1.1 where, for instance, the diagonal {1,3,13,85,781,...} is A023037 of Sloane; this opens other relations among the elements of the Fermatian array. We can also define a divisor step function $\delta(i,j)$ by

$$\delta(m,s) = \begin{cases} 1, & m \mid s, \\ 0, & m \nmid s, \end{cases}$$

When dealing with ordinary integers we note that has different interpretations in Hardy and Wright (1967) and Mollie Horadam (1966), namely

$$\begin{aligned} \delta(m,s) &= g(s,m)/m \\ &= \begin{cases} g(m)/m \\ E_s(m+1). \end{cases} \end{aligned}$$

We then have that

$$\theta(u_n, u_t) = \sum_{j=1}^{t} \delta((u_{n-j}, u_t), u_1)\delta((u_j, u_t), u_1) \tag{6.1.13}$$

Proof: The two delta functions in the summation are both unity only when $(u_{n-j},u_t)|u_1$ and when $(u_j,u_t)|u_1$; that is, only when

$$(u_{n-j}, u_t) = (u_j, u_t) = 1.$$

When we sum over all j up to and including t we have the number of elements which satisfy the conditions of Nagell's function for divisibility sequences. As a corollary we have

$$\varphi(u_n) = \theta(u_n, u_n) \qquad (6.1.14))$$

$$= \sum_{j=1}^{n} \delta((u_{n-j}, u_n), u_1) \delta((u_j, u_n, u_1))$$

since this yields the number of elements of the set $\{u_1, u_2, ..., u_{n-1}\}$ which are coprime with u_n.

Of course, there are still unsolved problems with some of the classical functions in terms of the ordinary integers (Wooldridge 1979). Sándor (1984) has also studied extensions of arithmetic functions. Analogues of other functions can be similarly defined; for example, the Nagell totient function for elements of a divisibility sequence. We define $\theta(u_n, u_t)$ as the number of elements u_j of the set $\{u_1, u_2, \ldots, u_t\}$, $t \leq n$, such that

$$(u_{n-j}, u_t) = (u_j, u_t) = 1.$$

Other possible future investigation would be to find an analogy to Wilson's theorem with the Fermatian numbers and to find a more comprehensive combinatorial description of them. For instance, Riordan (1968) has shown that

$$\underline{q}_{n+1} = 1 + a_{n1(q)} \qquad (6.1.15)$$

in which $a_{nm}(q)$ is the enumerator of partitions with m parts, none greater then n, such that their Ferrer's graphs include an initial triangle of sides n and m (the graph of the partition m, m-1, ... , 2, 1) (Macmahon 1915). Other visual perspective investigations include simple applications; for example,

- Shannon (2017) has applied the sequences in this section with Jacobian matrices and golden sections to food web modelling;
- Atanassov et al. (2002), Atanassov (2017) and Benjamin and Quin (2003) have utilized visual and combinatorial approaches to variations on recursive sequence themes.

6.2. SUMMARY

Rather than merely recapitulate the salient features of the preceding chapters, in this section we consider some aspects of the production of twin primes since, as we shall see, they encompass material from both strands of the monograph. This includes the sequence of Fibonacci numbers and the use of right-end-digits (REDs) which are independent of integer size so that relative comparisons with 'large' primes are readily made.

Integers may be expressed in the form nR where R is the right-end-digit (RED); for instance, 19157 has $R = 7$ and $n = 1915$. Table 6.2.1 illustrates this for certain values of n.

Table 6.2.1. Integer forms which yield primes

Form of n	R = 1	R = 3	R = 7	R = 9
n = 3t	X	---	X	---
n = 3t + 1	X	X	X	X
n = 3t + 2	---	X	---	X

Table 6.2.2a. ‡REDs with these *n* are always composite integers

n = f(t)	n = f(a,j)	‡	n = f(t)	n = f(a,j)	‡	n = f(t)	n = f(a,j)	‡
3t	all	3,9	3t + 1	21j + 4	9	3t + 2	all	1,7
	21j	3,7,9		21j + 7	7		21j + 11	9
	21j+9	1,3,9		21j + 13	3		21j + 20	3
				21j + 16	1			

Table 6.2.2b. Values of *t* which produce twin primes

RED type p_1^*, p_2^*	Examples	Prime sequences n_1, n_2	t
9,1	(29,31) (809,811) (3299,3301)	$n_1 = 3t_1 + 2$ $n_2 = 3t_2$	0,1,4,5,7,8,13,18,19,21,26,33,34,40,42,43,53, 64,70,76,77,84,90,92,99,103,109,...
7,9	(17,19) (857,859) (3557,3559)	$n_1 = 3t + 1$ $n_2 = 3t + 1$	0,3,4,6,7,11,15,20,27,28,42,47,49,53,55,56, 59,62,66,69,74,75,89,105,108,115,118,...
1,3	(11,13) (881,883) (2591,2593)	$n_1 = 3t + 1$ $n_2 = 3t + 1$	0,1,2,3,6,9,10,14,15,17,21,29,34,35,36,38,43, 48,49,57,62,64,69,70,71,79,86,...

For example, in an integer where $n = 3t = 21j$ with $R \varepsilon \{3,7,9\}$ such integers can never be primes; with $n = 3t + 1 = 21j + 7$ with $R = 7$ can never be prime (Table 6.2.2a).

On the other hand, Table 6.2.2b shows values of t which produce twin primes.

The three main sequences have imbedded sequences defined by

$$n = a + 21j \tag{6.2.1}$$

in which the value of a ranges from 0 to 20 but is restricted by the REDs of the primes of the main sequence in which they are imbedded (Table 6.2.3). Patterns, not otherwise apparent, start to emerge. The twin primes with REDs (1,3) and (7,9) have the simplest structure in $\{3t+1\}$, but the (1,3) primes are not formed if $a = 13$ or 16, and the (7,9) twins will not appear if $a = 4$ or 7 even though these primes also belong to the sequence $\{3t+1\}$ (Tables 6.2.3 and 6.2.4).

Table 6.2.3. Imbedded sequences

a	Main sequence	REDs yielding no primes	a	Main sequence	REDs yielding no primes	a	Main sequence	REDs yielding no primes
0	$3t$	3,9	7	$3t+1$	7	14	$3t+2$	1,7
1	$3t+1$	---	8	$3t+2$	1,7	15	$3t$	3,9
2	$3t+2$	1,7	9	$3t$	3,9	16	$3t+1$	1
3	$3t$	3,9	10	$3t+1$	---	17	$3t+2$	1,7
4	$3t+1$	9	11	$3t+2$	1,7	18	$3t$	3,9
5	$3t+2$	1,7	12	$3t$	3,9	19	$3t+1$	---
6	$3t$	3,9	13	$3t+1$	3	20	$3t+2$	1,7

Table 6.2.4. a **ε {13,16} in** $n = a + 21j = 3t + 1$

a	$R \neq 3$; a = 13		$R \neq 1$; a = 16	
	n	R	n	R
0	13	1,7,9	16	3,7
1	34	7,9	37	3,9
2	55	7	58	7
3	76	1,9	79	7
4	97	1,7	100	9
5	118	1,7	121	3,7
6	139	9	142	7,9
7	160	1,7,9	163	7
8	181	1	184	7
9	202	7,9	205	3
10	223	7,9	226	7,9

Twin primes with REDs (7,9): Table 6.2.5.

Twin primes with REDs (1,3): Table 6.2.6.

The first one hundred of these types of primes may be summarised in sequences that yield values for a and j.

Some examples for a = 1,4,10,19 are now provided in Table 6.2.7. Note the high prime yields.

Table 6.2.5. The first 100 twin primes with REDs (7, 9) from $n = a + 21j = 3t + 1$

a	j values which yield the primes	No. of twins
1	0,1,4,6,8,15,20,22,26,33,41,42,46,52,56,62,66,72,77,81,84,87	23
10	0,8,9,15,16,21,29,32,35,37,38,46,61,63,65,67,68,74,76,85	20
13	0,1,7,9,10,18,19,20,21,35,40,47,49,52,57,69,75,85	18
16	6,10,12,16,19,22,25,34,47,49,55,64,70,82,83	15
19	0,1,3,7,8,9,16,17,26,27,35,36,39,40,44,46,47,52,68,74,81,82,84,85	24

Table 6.2.6. The first 100 twin primes with REDs (1, 3) from $n = a + 21j = 3t + 1$

a	j values which yield the primes	No. of twins
1	0,2,3,5,6,7,10,16,17,19,21,43,45,48,49,54,60,61,67,76	20
4	0,2,3,4,5,6,8,9,10,18,20,26,29,31,32,39,42,44,49,53,65,66,68,74	24
7	0,111,12,13,18,20,21,27,32,34,56,57,58,65,66,73,78	18
10	0,1,2,5,15,16,17,19,22,23,39,49,57,69,74,79,80	17
19	0,3,4,6,8,9,12,23,24,25,26,31,34,42,44,48,57,61,74,77,85	21

Table 6.2.7. Examples of primes from $a = 1, 4, 10, 19$ from $n = a + 21j = 3t + 1$

j	a = 1		a = 4 (R ≠ 9)		a = 10		a = 19	
	n	p*	n	p*	n	p*	n	p*
0	1	1,3,7	1	1,3,7	10	1,3,7,9	19	1,3,7,9
1	22	3,7,9	25	1,7	31	1,3,7	40	1,9
2	43	1,3,9	46	1,3,7	52	1,3	61	3,7,9
3	64	1,3,7	67	3,7	73	3,9	82	1,3,7,9
4	85	3,7,9	88	1,3,7	94	1,7	103	1,3,9
5	106	1,3,9	109	1,3,7	115	1,3	124	9
6	127	7,9	130	1,3,7	136	1,7	145	1,3,9
7	148	1,3,7,9	151	1	157	1,9	166	3,7,9
8	169	3,7,9	172	1,3	178	3,7,9	187	1,3,7,9
9	190	1,7	193	1,3	199	3,7,9	208	1,3,7,9
10	211	1,3	214	1,3	220	3,7	229	3,7

Table 6.2.8. The first 100 twin primes with REDs (9, 1) from $n = a + 21$; $R^* = 9$, $n_1 = 3t_1+2$; $R^* = 1$, $n_2 = 3t_2$

a	j values which yield the primes	No. of twins
2,3	0,1,3,6,10,11,12,14,16,19,20,22,25,36,38,39,40,50,53,65,73,80	22
5,6	0,1,6,9,13,14,19,20,26,30,31,32,46,47,48,58,65,69,77,79,83,85,86,88	24
14,15	2,7,15,18,30,31,32,42,51,52,54,59,66,72,83,84	16
17,18	0,2,3,4,5,14,15,16,22,23,26,27,29,37,42,43,44,52,56,65,67,76,80,81,90	25
20,0	(10,11),(15,16),(27,28),(28,29),(29,30),(34,35),(41,42),(42,43),(49,50),(50,51),(54,55),(56,57)	13

Twin primes with REDs (9,1): These primes appear in different main sequences (Table 6.2.8).

When $a = 2$ and $a = 5$, $n = 3t + 2$, so that no primes with $R = 1$ or 7 can ever be produced. Similarly for $a = 3$ or 6, $n = 3t$, so that no primes with $R = 3$ or 9 will ever be produced in this sequence. Some examples are displayed in Table 6.2.9.

The integer structure can explain these results. There are also twin primes imbedded in the Fibonacci sequence. The Fibonacci primes only occur when the sequential (subscript) number is a prime and sometimes twin primes occur in this case (Table 6.2.10).

Extensions for generalizations of the Fibonacci numbers is a topic for further investigation. Recently, many number theoretic properties related to homogeneous linear recursive sequences have been made by many authors; see, for example, the papers by Ömür Deveci (2015; 2016; 2017).

Another useful analysis would be on the Brun's Constant, either using the directed structured approach (Equation (6.2.1)) or the RED structural method; that is, of finding the individual sums of the (1,3), (7,9) or (9,1) RED twin pairs.

Table 6.2.9. Examples of primes from *a* = 3,6 from *n* = 3*t* + 1 and *a* = 2,5 *n* = 3*t* + 2

j	a = 2		a = 3		a = 5		a = 6	
	n	p*	n	p*	n	p*	n	p*
0	2	3,9	3	1,7	5	3,9	6	1,7
1	23	3,9	24	1	26	3,9	27	1,7
2	44	3,9	45	7	47	9	48	7
3	65	3,9	66	1	68	3	69	1
4	86	3	87	7	89	---	90	7
5	107	---	108	7	110	3,9	111	7
6	128	3,9	129	1,7	131	9	132	1,7
7	149	3,9	150	---	152	---	153	1
8	170	9	171	---	173	3	174	1,7
9	191	3	192	---	194	9	195	1
10	212	9	213	1,7	215	3	216	1
11	233	3,9	234	1,7	236	---	237	1,7
12	254	3,9	255	1,7	257	9	258	---
13	275	3	276	7	278	9	279	1,7
14	296	3,9	297	1	299	9	300	1
15	317	9	318	1,7	320	3,9	321	7
16	338	9	339	1	341	3	342	---
17	359	3	360	7	362	3	363	1,7
18	380	3	381	---	383	3	384	7
19	401	3,9	402	1,7	404	9	405	1,7
20	422	9	423	1	425	3,9	426	1

Table 6.2.10. Some Fibonacci prime connections

p1	n1	p2	n2	Main sequences for p1, p2					
				{3t}		{3t+1}		{3t+2}	
3	0	5	0	X	X				
5	0	7	0	X	X				
11	1	13	1			X	X		
431	43	433	43			X	X		
569	56	571	57	X				X	

6.3. Concluding Comments

At a time when much attention is being focused on the STEM fields (Science, Technology, Engineering, Mathematics) even in primary or elementary education, there is a danger for the education bureaucrats of throwing the baby out with the bathwater, particularly when in most disciplines the first degree is no longer the "meal ticket". This underlines the importance of a strong broad and deep liberal foundation in prior education. This small monograph has in each section utilized the interplay of philosophy and history in mathematics with number theory as a vital part of the new *quadrivium.*

If the graduate attributes and learning outcomes of an institution's mission and vision are to be more than lip service in a rapidly changing world, there needs to be rigorous standards of scholarship at each stage of education, appropriate to the objectives of the institution, by

- encouraging research, scholarly and creative activities to influence our communities,
- contributing to a robust, equitable and environmentally sustainable society
- recognising and valuing diversity, and not imposing uniformity
- recognising the unique place of first peoples and their continuing contribution to the nation.

Here the Boyer (2016) model of scholarship is a worthy model with its message to tertiary education institutions of "don't try to be isomorphic with Harvard"!

But to put this situation right, it will not be enough merely to give more prominence to arts subjects (it is not the quantity that matters); it will be necessary to see the liberal arts – and the liberal sciences, as well – as parts of an integrated larger whole, capped by philosophy ("first philosophy"). Only in this way will it be possible to achieve what otherwise might seem impossible, namely "not growing old in the arts," but remaining true to the dreams of their youth, being truly young at heart

(McInerney, 1983). In the last analysis, immersion in the liberal arts in a connected way is a strong foundation for any profession and is the heart of the humanities, a means of living a full human life (White, 1998).

The cultivation of creativity requires inspiration from the teacher but it will not flourish if the approach is too simplistic or too mechanistic. A group of STEM teachers has adopted a "creative teaching" strategy to improve teacher morale and student retention (Pollard et al 2017).

All of this begs the question about who can teach in an integrated liberal arts program? Based on the foregoing such teachers should not only be scholars in one of the strands of the Liberal Arts but also be able to demonstrate an appreciation of the integration appropriate to a Liberal Arts program. The integration is much more than the casual references to links. The whole program has to be greater than the sum of the parts, in much the way that strands of string can produce a strong rope. It is not a matter of ticking the boxes as subjects are passed. This happens in *a la carte* liberal studies courses in some universities.

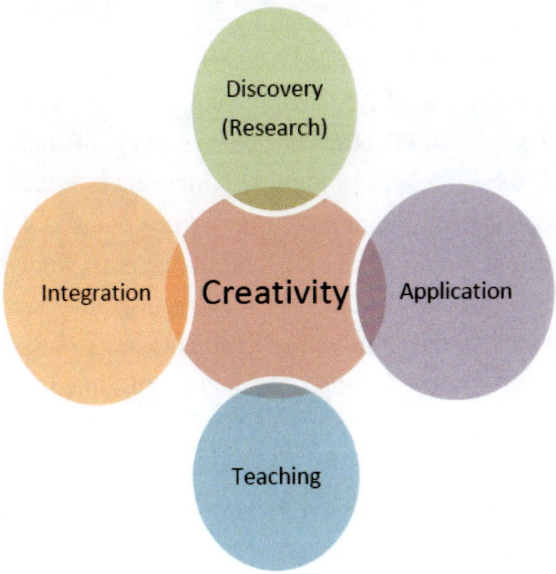

Figure 6.3.1. Boyer model of scholarship.

Metaphysics is at the heart of a liberal education that prepares the student for living. “Metaphysics is not just one among many subjects in philosophy. Unlike other fields, such as epistemology and aesthetics, metaphysics takes priority. Furthermore, metaphysics is a unique field of knowledge for culture and civilization. Arguably, the very identity and well-being of a civilization depends on whether it accepts metaphysics as a fundamental way of knowing… metaphysics, in its effort to grasp reality, is ultimately responsible for explaining how reason knows what is real” (Jaroszynski, 2011).

In the final analysis, Liberal Arts education is moral education – moral “in the most fundamental sense of forming the core of the personality, that underlying sense of direction and purpose from which the entire life’s pattern of thoughts, decisions and actions arise” (Walsh, *op. cit*). The intense moral vision of a single great artist has far more potential for changing hearts and minds than the pietistic clichés of a thousand preachers. For the artist speaks with the authority conveyed by suffering in the cause of truth, beauty and goodness and the realisation of one’s personal mediocrity and even nothingness, a recognition that inspires struggle to rise above all limitations.

But perhaps teachers of the Liberal Arts have not met so well their task of putting students in direct contact with the minds of such great men and women. Have they, perhaps, interposed themselves, their lecture notes, their textbooks, between their students and true genius? Of course, if there is a fault to be found it is not theirs alone. Since most students have access to few classic works, how could they read them if they are recommended to them? Sometimes, too, required assignments are so heavy that students have little time to “waste” with books that require meditation rather than memorization.

An even greater obstacle is the general lack of interest in today’s world in genuine reading. No one doubts the capacity of most students to “read” – even far into the night – so as to pass an impending examination. But librarians can attest that they rarely encounter university students engaged in serious leisure reading. Newspapers, paperback romances or thrillers, and spiritualistic morale-boosters probably constitute the main course, and even the only course, of any extracurricular reading students do. Walking city streets is hazardous because so many are being entertained by their "smart" phones!

What is to be deplored, however, is the narrowly pragmatic attitude which reduces all of education to the "training of manpower" in the service of the national economy. For that attitude loses sight of the priceless value of every unique individual; no person can live "by bread alone." Degrees are too often seen only as meal tickets for earning a living, not also as means for living.

But what do we understand by "liberal arts" and "liberal education"? And can it be defended in a time of serious economic recession? Is it, perhaps, at best a luxury that society cannot afford at the present time? Professor Christopher Wolfe of Marquette University defines liberal education as "an introduction to the ordered pursuit of the truth about reality" (Wolfe, 1985). It "seeks knowledge of the inter-related and integrated whole, as against the study of isolated bits of information about the whole." Its aim is "knowledge for its own sake, a good in itself (like health), something intrinsically desirable."

Peculiar to the liberal arts is their cathartic function: they help people to look at reality – and especially at themselves – in the face. This not only dispels illusions but also does the constructive service of inspiring magnanimity and many other virtues which are indispensable in one way or another to development to development as a person, and therefore as a citizen.

The liberal arts also make an effective and unique contribution to human communication. We do not need to be reminded how many official pronouncements and statements written by public functionaries are notorious for their ambiguity, vagueness, and banality. By contrast, we also know how one well-conceived cartoon can get more across than dozens of speeches. It is precisely because good art is such an effective communicator that there are circumstances when it is the best way to engage in social criticism and the best defence against tyranny. As Aleksandr Solzhenitsyn has said, "Wherever else it fails, Art always has won its fight against lies, and it will always win... The lie can withstand a great deal in this world but in cannot withstand Art" (Solzhenitsyn, 1972).

Nevertheless, if the arts undoubtedly express values and make them shine, they cannot create or justify those values. The great artist is simply the one who can see more deeply into reality than the rest of us, who can catch some hidden truth, and successfully translate it into understandable language. The arts are not an end in themselves; they naturally point us toward a realm above them, to philosophy and its pursuit of the ultimate causes of all things. It is the "first philosophy" (metaphysics) which leads us to the foundation of those values which the liberal arts explain and illustrate.

Metaphysics provides the rational basis and the ordering scheme by which all of the arts and sciences can be understood in their rightful internal autonomy. It does this by clarifying the first principles which each discipline takes for granted as it pursues its proper objects (Maritain, 1932). Here they have the reason why Aristotle reserved for metaphysics the title of Wisdom. This was later expanded by the medieval thinkers into the maxim, "non est senescendum in artibus" (one is not to grow old in the arts). If, as they thought, the beginning of wisdom is to be found in the liberal arts of the trivium and the quadrivium, Wisdom itself was only to be reached (on the level of human knowledge) in metaphysics.

The discipline of philosophy enjoys so little esteem and is understood so poorly that most people will laugh at the assertion that wisdom is only to be found within it. It cannot be surprising that philosophy invites little respect when the Liberal Arts in general are not regarded as very serious subjects. But if this matter is so important, so fraught with consequences, we must try to pursue the reasons for such superficial views. Why have education specialists and educational reformers failed to appreciate the supremely formational role of a liberal education capped by philosophy? Why do so many supposedly well-educated people resort instead to "spiritualistic" sects based principally on sentiment?

History can help us here by contrasting the esteem in which the liberal arts were held in remote times and places. Raymond Klibansky is referring in the following passage from the early medieval School of Chartres: "The seven liberal arts together give man both knowledge of the divine and the power... to express it. But in doing so, they fulfil at the same time another purpose. They serve ad cultum humanitatis, that is, they promote the specifically human values, revealing to man his place in the universe and teaching him to appreciate the beauty of the created world" (Wagner, 1983).

And what was the "humanism" which the medievalist Richard William Southern considered not only typical of the period from 1100 to 1320 but even as the catalyst of what later came to be known as the Renaissance? That humanism connoted for Southern an emphasis on human dignity and reason, a recognition of the order and intelligibility of nature (Southern, 1970). Thus, in a sense, the work of Leonardo Fibonacci of Pisa came to be better known in Europe in the nineteenth century in a general way with Libri (1838) and more mathematically with Lucas (1878).

Sometimes historians are accused of advocating a return to the past. That, of course, is neither possible nor desirable (Maritain, 1938). But one can advocate the recovery of an attitude which is open to past achievements and in particular to the understanding of wisdom as the most perfect, the most noble, the most useful, and the most joyful of all human pursuits.

In a pragmatic and relativistic society such as ours, the effort to restore metaphysics and a realist philosophy in general might seem doomed from the start. Nevertheless, one who is disposed to make the attempt can surely gain inspiration from the reflection that reformers always have to swim against the tide of contemporary fashion and prejudice. A remarkable example of upstream swimming is provided by Robert Maynard Hutchins, President of the University of Chicago between 1929 and 1951, and an advocate with Mortimer Jerome Adler of "The Great Books of Western Civilization" (Adler, 1940). A few words on Hutchins can serve to emphasise these points (Neg, 1959).

Soon after becoming President of the University of Chicago, Hutchins took the highly unorthodox and unusual decision "to leave undone a vast number of things that university presidents can do" in order to dedicate himself to self-education and research. He had become convinced that despite being a graduate of Yale University and Yale Law School he was still "uneducated," and that "as a man to whom had been confided one of the major educational posts in the country, he should begin by trying to get an education." He thus set about educating himself "with the help of some great books that have endured, and with twenty to thirty undergraduates as fellow students, selected from each incoming class for their promise." This was the beginning of a four-year course in the Great Books.

This self-imposed course enabled him to bring about within the University of Chicago an intellectual and educational revolution that would make that university one of the best, not only in the United States but in the world. (Gress has a slightly different view on the Great Books approach to Western Civilisation which is worth noting but it would be too much of a digression to pursue here (Gress, 1988).) Hutchins stressed that the great works of the distant past provide a scale of values against which to judge the activities and educational offerings of his own age, and that these older values have an intellectual and scientific solidity which the contemporary pragmatist and relativist mentality can hardly imagine. He believed that metaphysics, as conceived by Aristotle, can help people have the foundations for the acquisition of wisdom in ways that applied studies, which are concerned with the particular, cannot.

The Golden section permeates this book Informally it is also a common thread through the ancient Pythagorean quadrivium to which we have previously referred:

- Arithmetic – pure number,
- Geometry – number in space,
- Spherics – number in space and time,
- Music – number in time.

The word 'music' (μουσικ'η) had a wider meaning in ancient Greece than it does now. It embraced "the idea of ratios of integers as the key to understanding both the visible physical universe and the invisible spiritual universe". (Benson, 2006).

Henri Bergson once said, “The body, now larger, calls for a bigger soul.” This remark seems very relevant to education in the Western world today. The material structure of education is now much larger and more elaborate than it was when we were undergraduates. But our material growth has not been matched by a corresponding spiritual growth. We assert that one of the reasons for this is the neglect of the liberal arts and failure to delve deeply into the purpose of education which has characterized educational planning in recent years. Yet government reviews of education deal with the ‘what’ always, the ‘when’ and the ‘how’ occasionally, but the ‘why’ almost never. This was a point made by the late great Bill Radford, Director of the Australian Council for Educational Research, who observed wryly that Australians had a roll-up-one’s-sleeves approach to the philosophy of education! (Radford, 1964). Things have not changed to judge from recent reports.

BIBLIOGRAPHY

Adler, Irving. 2006. "Nonegenarian Fibonacci Devotee." *The Mathematical Intelligencer*, 28 (1): 4.

Adler, Irving. 1969. "Three Diophantine Equations – Part II." *The Fibonacci Quarterly*. 7 (3): 181–193.

Adler, Mortimer J. 1940. *How to Read a Book: The Art of Getting a Liberal Education*, New York: Simon and Schuster.

Aküzüm and O. Deveci. 2017. "Sylvester-Padovan-Jacobsthal-type sequences." *Maejo International Journal of Science and Technology*. 11 (3): 236-248.

Aküzüm, Y., O. Deveci and A. G. Shannon. 2017. "On the Pell p-circulant sequences." *Notes on Number Theory and Discrete Mathematics*. 23 (2): 91-103.

Anatriello, G., and G. Vincenzi. 2014. "Tribonacci-like Sequences and Generalized Pascal's Pyramids." *International Journal of Mathematical Education in Science and Technology*. 45 (8): 1220–1232.

Apostol, T. M. 1980. *Introduction to Analytic Number Theory*. Berlin: Springer-Verlag.

Araci, S., M. Acikgoz, and E. Şen. 2013. "On the Extended Kim's *p*-adic *q*-deformed Fermionic Integrals in the *p*-adic Integer Ring." *Journal of Number Theory*. 133 (1): 3348–3361.

Atanassov, Krassimir T. 2017. *Intutionistic Fuzzy Logics*. Berlin: Springer-Verlag.

Atanassov, Krassimir T. 2017. "On two new two-dimensional extensions of the Fibonacci sequence." *Notes on Number Theory and Discrete Mathematics*. 23 (3): 115-122.

Atanassov, Krassimir T., Daryl R. Deford, and Anthony G. Shannon. 2015. "Pulsated Fibonacci Recurrences." In Peter G. Anderson, Christian Ballot, William Webb (eds), *The Fibonacci Quarterly*. 52 (5): 22-27.

Atanassov, K., V. Atanassova, A. Shannon, and J. Turner. 2002. *New Visual Perspectives on the Fibonacci Numbers*. New York: World Scientific.

Atanassov, K. T., and A. G. Shannon. 1998. "A note on intuitionistic fuzzy logics." *Acta Philosophica*. 7 (1):121-125.

Atanassov, K. T., and A. G. Shannon. 2001. "The Digital Root Function for Fibonacci-type Sequences." *Advanced Studies in Contemporary Mathematics*. 21 (3): 251–254.

Atanassov, K. T., and A. G. Shannon. 2008. "A Fibonacci Cylinder." *Notes on Number Theory & Discrete Mathematics*. 14 (4): 4-9.

Atanassov, K. T., A. G. Shannon and J. C. Turner. 1997. "Remark on Fibonacci sequences and fuzzy sets." *Comptes rendus de l'Academie Bulgare des Sciences*. 50 (2): 11-14.

Baker, Alan *A*. 1984. *Concise Introduction to the Theory of Numbers*. Cambridge: Cambridge University Press.

Baker, A. B. Bollabás, and A. Hanjel (eds). 1990. *A Tribute to Paul Erdős*. Cambridge: Cambridge University Press.

Ball, W. W. R., and H. S. M. Coxeter. 1956. *Mathematical Recreations and Essays*. New York, Macmillan.

Ballot, Christian. 2017. "Divisibility of the Middle Lucasnomial Coefficient." *The Fibonacci Quarterly*. 55 (4): 297-308.

Belaga, E., and M. Mignotte. 1998. "Embedding the 3x+1 Conjecture in a 3x+d Context." *Experimental Mathematics*. 7(2): 145-151.

Benjamin, A. T. 2007. *The Joy of Mathematics*. Chantilly, VA: The Great Courses.

Benjamin, Arthur T., and Jennifer J. Quinn. 2003. *Proofs That Really Count: The Art of Combinatorial Proof.* Washington DC: Mathematical Association of America.

Bennett, A. A. 1925. "Two New Arctangent Relations for π." *The American Mathematical Monthly*. 32: 253–255.

Besnon, David I. 2006. *Music: A Mathematical Offering.* Cambridge: Cambridge University Press, p.193.

Berenhaut, K. S., A. B. O'Keefe, and F. Saidak, F. 2010. "Remarks on Linear Recurrences of the Form $y_n = y_{n-1} + a_{n-1}y_{n-2}$." *Congressus Numerantium*. 200, 141–151.

Berge, C. 1971. *Principles of Combinatorics.* New York: Academic Press, 1971.

Berndt, B., and S. Bhargava. 1993. "Ramanujan-for lowbrows." *American Mathematical Monthly*. 100: 645-656.

Birch, R. H. 1946. "An Algorithm for the Construction of Arctangent Relations." *Journal of the London Mathematical Society.* 21 (2): 173–174.

Bollabás, A. Hanjel (eds). 1990. *A Tribute to Paul Erdős*. Cambridge: Cambridge University Press, pp.105-110.

Boncompagni-Ludovisi, Baldassarre. 1854. *Intorno ad alcune opere di Leonardo Pisano.* Roma: Atti dell'Accademia Pontificia dei Nuovi lincei: tipografia delle belle arti. [*Around some works* by Leonardo Pisano. Rome: Proceedings of the Pontifical Academy of the New Lincei: typography of the fine arts.]

Bondarenko, Boris A. 1993. *Generalized Pascal Triangles and Pyramids: Their Fractals, Graphs, and Applications.* (Translated by Richard C Bollinger.) Santa Clara, CA: The Fibonacci Association.

Bourgain, A. 1990. "On arithmetic progressions in sums of sets of integers." In *A Tribute to Paul Erdős* edited by Baker, A. B. Bollabás, and A. Hanjel. Cambridge: Cambridge University Press, pp.105-110.

Brent, Richard P., and Graeme L. Cohen. 1989. "A new lower bound for perfect numbers." *Mathematics of Computation*. 53 (187): 431- 437.

Bruckman, Paul S. 2008. "Erratum: A proof of the Collatz conjecture." *International Journal of Mathematical Education in Science and Technology.*39 (4): 567.

Carlitz, L. 1940. "A Set of Polynomials." *Duke Mathematical Journal.* 6 (3): 486-504.

Carlitz, L. 1948. "q-Bernoulli numbers and polynomials." *Duke Mathematical Journal.* 15 (6): 987-1000.

Carlitz, L. 1954. "Congruences for the solutions of certain difference equations of the second order." *Duke Mathematical Journal.* 21 (4); 669-679.

Carlitz, L. 1964a. "A functional-difference equation." *Duke Mathematical Journal.* 31 (4): 667-690.

Carlitz, L. 1964b. "A binomial Identity Arising from a Sorting Problem." *SIAM Review.* 6 (1): 20–30.

Carlitz, L., and L. Moser. 1966. "On some special factorizations of $(1–xn)/(1–x)$." *Canadian Mathematical Bulletin.* 9 (4) 421-426.

Childs, Lindsay. 1979. *A Concrete Introduction to Higher Algebra.* New York: Springer, p. 219.

Clarke, J. H., A. G. Shannon and J. V. Leyendekkers. 1998. "On sums of pairs of squares and cubes." *Notes on Number Theory and Discrete Mathematics.* 4 (3): 108-111.

Cohen, G. L. 1975. "Selberg formulae for Gaussian integers." *Acta Arithmetica.* 16: 385-400.

Collatz, Lothar 1986. "On the motivation and origin of the (3n+1) problem." (In Chinese). *Journal of Qufu Normal University, Natural Science Edition.* 3: 9-11. Translated and reproduced in Jeffrey C. Lagarias (ed.) 2010. *The Ultimate Challenge: the 3*x*+1 problem.* Providence, RI: the American Mathematical Society.

Conway, J. H. 1972. "Unpredictable iterations." *Proceedings of a Number Theory Conference.* Boulder, CO: University of Colorado, pp. 49-52.

Conway, J. H., and R. K. Guy. 1996. *The Book of Numbers*. New York: Copernicus.

Cook, C. K., and A. G. Shannon. 2006. “Generalized Fibonacci and Lucas Sequences with Pascal-type Arrays.” *Notes on Number Theory and Discrete Mathematics*. 12 (4): 1–9.

Cox, N., J. W. Phillips, and Verner E. Hoggatt, Jr. 1970. “Some universal counterexamples.” *The Fibonacci Quarterly*. 8 (2): 242-248.

Coxeter, H. S. M. 1953 “The Golden Section, Phyllotaxis, and Wythoff's Game.” *Scripta Mathematica*. 19, 135–143.

Crăciun, I., D. Invan, and D. Popa. 2015. “Generalized Golden Ratios defined by means.” *Applied Mathematics and Computation*. 250 (1): 221–227.

Daykin, David E. 1963. “Linear sequences over a finite field.” *American Mathematical Monthly*. 70: 637-642.

Deakin, Michael. A. B. 2013. “Theano: the World's First Female Mathematician?” *International Journal of Mathematical Education in Science and Technology*. 44 (3): 350–364.

Deveci, O. 2016. “The Pell-circulant sequences and their applications.” *Maejo International Journal of Science and Technology*. 10 (3): 284-293.

Deveci, O. and E. Karaduman. 2015. “The Jacobsthal-circulant sequences and their applications.” *ROMAI Journal*. 11 (2): 77-88.

Deveci, O., E. Karaduman and C. M. Campbell. 2017. “The Fibonacci–circulant sequences and their applications” *Iranian Journal of Science and Technology, Transactions A: Science*. https://doi.org/10.1007/s40995-017-0317-7.

Deveci, O. and A. G. Shannon. 2017. “Pell–Padovan-circulant sequences and their applications.” *Notes on Number Theory and Discrete Mathematics*. 23(3): 100-114.

Dewey, John. 1916. *Democracy and Education*. New York: Macmillan.

Dolgy, D. V., Taekyun Kim, and Jong Jin Seo. 2016. “Symmetry identities for Changhee polynomials arising from the Fermionic p-adic integral on Z_p.” *Advanced Studies in Contemporary Mathematics*. 26 (2): 291-298.

Drobot, V. 2000. “On Primes in the Fibonacci sequence.” *The Fibonacci Quarterly*. 38 (1),: 71–72.

Edwards, H. M. 1974. *Riemann's Zeta Function*. New York: Academic Press.

Edwards, Harold M. 1977. *Fermat's Last Theorem: A Genetic Introduction to Algebraic Number Theory, Graduate Texts in Mathematics, 50. New York: Springer, pp. 61–65*

Effinger, G., K. Hicks, and G. L. Mullen. 2005. "Integers and Polynomials: Comparing the Close Cousins Z and $F_q[x]$." *The Mathematical Intelligencer* 27 (2): 26-34.

Emerson, E. I. 1969. "Recurrent Sequences in the Equation DQ^2 = R^2 + N." *The Fibonacci Quarterly*. 7 (3): 231–242.

Erdös, P., and E. Jabotinsky. 1958. On Sequences of Integers Generated by a Sieving Process. *Nederlandse Akademie van Wetenschappen. Series A*. 61: 115–128.

Ferns, H. H. 1969. Products of Fibonacci and Lucas Numbers. *The Fibonacci Quarterly*. 7: 1-13.

Fielder, D. C. 1968. "Special Integer Sequence Controlled by Three Parameters." *The Fibonacci Quarterly*. 6 (1): 64–70.

Filipponi, P. 1991. "A Note on a Class of Lucas Sequences." *The Fibonacci Quarterly*. 29 (3): 256–263.

Forder, H. G. 1963. "A Simple Proof of a Result on Diophantine Approximations." *The Mathematical Gazette*. 47: 237–238.

Franklin, J. 2014. *An Aristotelian Realist Philosophy of Mathematics: Mathematics as the Science of Quantity and Structure*. New York, Palgrave Macmillan, p. 140.

Franklin, J. 2016. "Logical probability and the strength of mathematical conjectures." *Mathematical Intelligencer*. 38 (3): 14-19.

Fujiwara, M., and Y. Ogawa. 2005. *Introduction to Truly Beautiful Mathematics*. Tokyo: Chikuma Shobo.

Gardiner, A. 1982. "Digital Roots, Rings and Clock Arithmetic." *The Mathematical Gazette*. 66 (437): 184–188.

Gardner, Howard. 2011. *Truth, Beauty, and Goodness Reframed: Educating for the Virtues in the Twenty-First Century*. New York: Basic Books (Perseus Books Group).

Girard, Albert (ed.). 1618. *Oeuvres Mathematiques de Simon Stevin.* [Mathematical Works of Simón Stevin.]Leiden: Chez Bonaventure and Abraham Elsevier.

Gougenbaum, A. 1971. "About the Linear Sequence of Integers and that Each Term is the Sum of the Two Preceding." *The Fibonacci Quarterly*. 9 (3): 277–295, 298.

Gould, H. W. 1969. "The Bracket-function and Fontené-Ward generalized binomial coefficients with applications to Fibonomial coefficients." *The Fibonacci Quarterly*. 9 (1): 23-40,55.

Grabner, P. J., T. Herendi, and R. F. Tichy. 1997. "Fractal Digital Sums and Codes." *Applicable Algebra in Engineering, Communication and Computing*, 8 (1): 33–39.

Gress, David. 1998. *From Plato to NATO*. New York: Free Press (Simon and Schuster).

Griffiths, M. 2010. "Digit Proportions in Zeckendorf Representations." *The Fibonacci Quarterly*. 48 (2): 168–174.

Gutiérrez, Rochelle. 2018. "Political Conocimiento for Teaching Mathematics: why Teachers Need it and How to develop it." In *Building Support for Scholarly Practices in Mathematics Methods,* edited by Signe E. Kastberg, Andrew M. Tyminski, Alyson E. Lischka and Wendy B. Sanchez. . Charlotte, NC: Information Age Publishing, pp. 11-37.

Guy, Richard K. 1983. "Don't try to solve these problems." *American Mathematical Monthly*. 90 (1): 35-41.

Guy, R. K. 1994. "Sums of like powers. Euler's Conjecture. *Unsolved Problems in Number Theory.* 2nd edition. New York: Springer-Verlag, pp.139-144.

Guy, Richard K. 2004. *Unsolved Problems in Number Theory. Third edition.* New York: Springer, pp. 336-337.

Halls, Rachel Wells. 2008. *The Sound of Numbers: A Tour of Mathematical Music Theory.* Philadelphia, PA: St Joseph's University, pp. 8-18.

Halmos, P. R. 1968. "Mathematics as a Creative Art." *American Scientist*. 56 (4): 375-389.

Halmos, Paul. 1985. *I want to be a Mathematician.* Washington DC: Mathematical Association of America.

Hardy, G. H. 1937. "The Indian Mathematician Ramanujan." *American Mathematical Monthly*. 44 (3): 137-155.

Hardy, G. H., and Edward M. Wright. 1938. revised by D. R. Heath-Brown, J. H. Silverman, Andrew Wiles. (2008). *An Introduction to the Theory of Numbers* (*6th edition*). Oxford: Oxford University Press.

Hardy, G. H. 1940. *A Mathematician's Apology.* Cambridge: Cambridge University Press.

Hardy, G. H., and E. M. Wright. 1965. *An Introduction to the Theory of Numbers. 4th edition.* Oxford: Clarendon Press.

Havil, J. 2012. *The Irrationals: A Story of the Numbers You Can't Count on*. Princeton and Oxford: Princeton University Press, pp.5, 74.

Hermite, C. 1859. "Sur la théorie des équations modulaires." ["On the theory of modular equations."] *Comptes Rendus de l'Académie des Sciences, Paris*. 49: 16-24, 110-118, 141-144.

Heubach, Silvia, and Toufik Mansour. 2009. *Combinatorics of Compositions and Words.* Boca Raton, FL: Chapman and Hall/CRC.

Hilton, A. J. W. 1974. "On Ferns' Theorem on the Expansion of Fibonacci and Lucas Numbers." *The Fibonacci Quarterly.* 12: 231-232.

Hillman, A. P, and G. L. Alexanderson. 1973. *A First Undergraduate Course in Abstract Algebra.* Belmont: Wadsworth.

Hilton, P., and J. Pedersen. 2012. "Mathematics, Models, and Magz. Part 1: Patterns in Pascal's Triangle and Tetrahedron." *Mathematics Magazine.* 85 (2): 79–109.

Hoffman, P. 1998. *The Man Who Loved Only Numbers: The Story of Paul Erdős and the Search for Mathematical Truth.* New York: Hyperion, p. 190.

Hoggatt, V. E. Jr. 1967. Fibonacci numbers and generalized binomial coefficients. *The Fibonacci Quarterly*. 5 (4): 383-400.

Hoggatt, V. E. Jr. 1968. "A New Angle on Pascal's Triangle." *The Fibonacci Quarterly.* 6 (4): 221–234.

Hoggatt, V. E. Jr. 1969. *Fibonacci and Lucas Numbers.* Boston, MA: Houghton Mifflin.

Hoggatt, Verner. 1983. *Αριθμοι Fibonacci και Lucas*. (Μεταφραση: Ανδρεασ Ν. Φιλιππου & Γιωργοσ Ν. Φιλιππου). Gutenberg.

Hoggatt, Verner E. Jr, and Gerald L. Alexanderson. 1971. "A property of multinomial coefficients." *The Fibonacci Quarterly*. 9 (4): 351-356, 420-421.

Hoggatt, Verner E. Jr., Marjorie Bicknell. 1969. Diagonal sums of generalized Pascal triangles. *The Fibonacci Quarterly*. 7: 341-358.

Hoggatt, V. E. Jr., M. Bicknell-Johnson. 1977. "Fibonacci Convolution Sequences." *The Fibonacci Quarterly*. 15 (2); 117–122.

Hoggatt Jr, V. E. Jr, and M. Bicknell-Johnson (eds). 1980. *A Collection of Manuscripts related to the Fibonacci Sequence*. Santa Clara: The Fibonacci Association, 1980, pp.93–94.

Hoggatt, V. E. Jr, and D. A. Lind. 1968. "Fibonacci and Binomial Properties of Weighted Compositions." *Journal of Combinatorial Theory*. 4: 121-124.

Hoggatt, V. E. Jr, and D. A. Lind. 1969. Compositions and Fibonacci Numbers. *The Fibonacci Quarterly*. 7: 253-266.

Hongquan, Yu, Yi Wang and Mingfeng He. 1996. "On the limit of generalized Golden numbers." *The Fibonacci Quarterly*. 34 (4): 320–322.

Horadam, A. F. 1961. "Fibonacci number triples." *American Mathematical Monthly*. 68: 455–459.

Horadam, A. F. 1965a. Basic properties of a certain generalized sequence of numbers. *The Fibonacci Quarterly*. 3 (3): 161-176.

Horadam, A. F. 1965b. Generating functions for powers of a certain generalized sequence of numbers. *Duke Mathematical Journal*. 32 (4): 437-446.

Horadam, A. F. 1966 "Generalization of Two Theorems of K. Subba Rao." *Bulletin of the Calcutta Mathematical Society*. 58 (1): 23–29.

Horadam, A. F. 1967. Special properties of the sequence $\{w_n(a,b;p,q)\}$. *The Fibonacci Quarterly*. 5 (5): 424-434.

Horadam, A. F. 1968. Generalizations of two theorems of K. Subba Rao. *Bulletin of the Calcutta Mathematical Society*. 58 (1): 23-29.

Horadam, A. F. 1971. "Pell Identities." *The Fibonacci Quarterly*. 9 (3): 245–252.

Horadam, A. F., and A. G. Shannon. 1988. "Asveld's polynomials." In, *Applications of Fibonacci Numbers, Volume 3,* edited by A. N. Philippou, A. F. Horadam and G. E. Bergum. Dordrecht, Kluwer, pp. 163–176.

Horadam, A. F., and A. G. Shannon. "Pell-type number generators of Pythagorean triples." In *Applications of Fibonacci Numbers, Volume 5,* edited by G. E. Bergum, A. N. Philippou and A. F. Horadam Vol. 5, 1993, 331–343.

Horadam, E. Mollie. 1966. "Exponential functions for arithmetical semi-groups." *Journal für die reine und angewandte Mathematik.* 222: 14-19.

Horadam, E. Mollie. 1971. An extension of Daykin's generalized Möbius function to unitary divisors. *Journal für die reine und angewandte Mathematik.* 246: 117-125.

Ireland, Kenneth, and Michael Rosen. 1990. A Classical Introduction to Modern Number Theory. *Graduate Texts in Mathematics, 84*. New York: Springer, Ch.17.

Iverson, K. E. 1980. "Notation as a tool of thought." *Communications of the Association for Computing Machinery*. 23: 444-465.

Jarden, D. 1966. *Recurring Sequences*. Jerusalem: Riveon Lematematika.

Jaroszynski, Piotr. 2011. "Aquinas Metaphysics: Versus Mythology, Ontology, and Ideology." *Thomism and Asian Cultures: Quadricentennial International Philosophy Conference.* Manila: University of Santo Tomas, pp.86-98.

Jerbic, Stephen K. 1968. *Fibonomial Coefficients.* Master of Arts Thesis. San Jose State College, California.

Kaneda, Masayoshi. 2015. "Two remarks on the Collatz cycle conjecture." *The Fibonacci Quarterly.* 53 (2): 168-174.

Kanigel, R. 1991. *The Man Who Knew Infinity: A Life of the Genius Ramanujan.* New York: Washingon Square Press.

Kapur, J. N. 1976. "Generalized Pascal's Triangles and Generalized Fibonacci Numbers." *Pure and Applied Mathematical Sciences*. 3(1–2), 93–100.

Kapur, J. N. 1988. "Some generalizations of the golden ratio." *International Journal of Mathematical Education in Science and Technology*. 19 (4): 511–517.

Kasner, E., and J. Newman. 1959. *Mathematics and the Imagination*. Norwich: Jarrold.

Kim, D. S., T. Kim, and H. Y. Lee. 2013. "p-adic q-integral on Z_p associated with Frobenius-type Eulerian Polynomials and Umbral Calculus." *Advanced Studies in Contemporary Mathematics*. 23 (2): 243–251.

Kim, T., L. C. Jang, and C. Adiga. In press. "A Note on q-Euler Numbers." *International Journal of Mathematical Education in Science and Technology*.

Knopp, K. 1990. *Theory and Application of Infinite Series*. New York, Dover.

Knott, R. 2015. "*Fibonacci Numbers and the Golden Section.*" Available online: http://www.maths.surrey.ac.uk/hosted-sites/R.Knott/Fibonacci/fib.html.

Knuth, D. E. 2005. *Art of Computer Programming, Volume 4*. New York: Addison-Wesley, p. 50.

Krall, H. L. 1960. "Self-adjoint Differential Expressions." *American Mathematical Monthly*. 67: 876-878.

Krishna, H. V. 1974. "Pythagorean triads." *The Mathematics Student*. 47 (1): 41–43.

Lafer, Phil. 1971. "Discovering the Square-triangular Numbers." *The Fibonacci Quarterly*. 9 (1): 93-105.

Lagarias, Jeffrey C. 1985. "The $3x+1$ problem and its generalizations." *American Mathematical Monthly*. 92 (1): 3-23.

Lagarias, Jeffrey C. (ed.) 2010. *The Ultimate Challenge: the 3x+1 problem*. Providence, RI: the American Mathematical Society.

Larcombe, Peter J. 2017. "Mathematics as a Mirror of Painting." *Mathematics Today*. December, pp.283-285.

Larcombe, Peter J. 2018. "A few thoughts on the aesthetics of mathematics in research and teaching." *Palestine Journal of Mathematics*. 7 (1): 1-8.

Larsen, R. J., Marx, M. L. 2001. *An Introduction to Mathematical Statistics and Its Applications.* Upper Saddle River, NJ: Prentice Hall, p. 282.

Leahy, D. G. 1996. *Foundation: Matter the Body Itself.* NY: State University of New York Press, p.523.

Lehmer, D. H. 1935. "A Cotangent Analogue of Continued Fractions." *Duke Mathematical Journal.* 4: 323–340.

Lehmer, D. H. 1936. "Problem 3801." *The American Mathematical Monthly.* 43: 580.

Lehmer, D. H. 1938. "On Arctangent Relations for π." *The American Mathematical Monthly.* 45: 657–664.

LeVeque, W. J. 1977. *Fundamentals of Number Theory.* Reading, MA: Addison-Wesley.

Leyendekkers, J. V., and J. M. Rybak. 1995. "Pellian Sequences Derived from Pythagorean Triples." *International Journal of Mathematical Education in Science and Technology*. 26 (6): 903–922.

Leyendekkers, J. V., and J. M. Rybak. 1995. "The Generation and Analysis of Pythagorean Triples within a Two-parameter Grid." *International Journal of Mathematical Education in Science and Technology*. 26 (5); 787–793.

Leyendekkers, J. V., J. M. Rybak, and A. G. Shannon. 1995. "Integer class properties associated with an integer matrix." *Notes on Number Theory and Discrete Mathematics*. 1 (2): 53–59.

Leyendekkers, J. V., J. M. Rybak, and A. G. Shannon. 1997. "Analysis of Diophantine Properties Using Modular Rings with Four and Six Classes." *Notes on Number Theory and Discrete Mathematics*. 3 (2): 61-74.

Leyendekkers, J. V., and A. G. Shannon. 1998. "Fibonacci Numbers within Modular Rings." *Notes on Number Theory and Discrete Mathematics*. 4 (4): 165–174.

Leyendekkers, J. V., and A. G. Shannon. 1999. "Analysis of Row Expansions within the Octic 'Chess' Modular Ring, Z_8." *Notes on Number Theory and Discrete Mathematics*. 5 (3): 102-114.

Leyendekkers, J. V., and A. G. Shannon. 2001. "The Analysis of Twin Primes within Z_6." *Notes on Number Theory and Discrete Mathematics*. 7 (4): 115-124

Leyendekkers, J. V., and A. G. Shannon. 2002a. "Powers as Differences of Squares: The Effect on Triples." *Notes on Number Theory and Discrete Mathematics,* 8 (2): 95-106.

Leyendekkers, J. V., and A. G. Shannon. 2002b. "Constraints on powers within the modular ring Z_4: Part 1." *Notes on Number Theory and Discrete Mathematics,* 8 (2): 41-54.

Leyendekkers, J. V., and A. G. Shannon. 2002c. "Constraints on powers within the modular ring Z_4: Part 2." *Notes on Number Theory and Discrete Mathematics,* 8 (2): 58-66.

Leyendekkers, J. V., and A. G. Shannon. 2002d. "Constraints on powers within the modular ring Z_4: Part 3." *Notes on Number Theory and Discrete Mathematics,* 8 (3): 95-106.

Leyendekkers, J. V., and A. G. Shannon. 2003. "Some Characteristics of Primes within Modular Rings." *Notes on Number Theory and Discrete Mathematics,* 9 (1): 49-58.

Leyendekkers, J. V., and A. G. Shannon. 2004a. "The Row Structure of Squares in Modular Rings." *Notes on Number Theory and Discrete Mathematics*. 10 (1): 1-11

Leyendekkers, J. V., and A. G. Shannon. 2004b. "The Structure of Fibonacci Numbers in Modular Rings." *Notes on Number Theory and Discrete Mathematics*. 10 (1): 12-23.

Leyendekkers, J. V., and A. G. Shannon. 2004c. "Using Integer Structure to Calculate the Number of Primes in a Given Interval." *Notes on Number Theory and Discrete Mathematics*. 10 (3): 77-83.

Leyendekkers, J. V., and A. G. Shannon. 2005. "Infinite Series and Modular Rings." *Notes on Number Theory and Discrete Mathematics,* 11 (3): 7-14.

Leyendekkers, JV, and AG Shannon. 2007a. "Odd powers as a sum of squares." *Notes on Number Theory and Discrete Mathematics*. 13 (1): 16-24.

Leyendekkers, JV, and AG Shannon. 2007b. "Integer Structure Analysis of Primes and Composites from Sums of Two Fourth Powers." *Notes on Number Theory and Discrete Mathematics*. 13 (2): 1-9.

Leyendekkers, J. V., and A. G. Shannon. 2008a. "Analysis of Primes Using Right-end-digits and Integer Structure." *Notes on Number Theory and Discrete Mathematics.* 14(3): 1-10.

Leyendekkers, J. V., and A. G. Shannon. 2008b. "The identification of Rows of Primes in the Modular Ring Z_6." *Notes on Number Theory and Discrete Mathematics.* 14(3): 10-16.

Leyendekkers, J. V., and A. G. Shannon. 2010. "Equations for Primes obtained from Integer Structure." *Notes on Number Theory and Discrete Mathematics.* 16 (3): 1-10.

Leyendekkers, J. V., and A. G. Shannon. 2011a. "Modular Rings and the Integer 3." *Notes on Number Theory and Discrete Mathematics.* 17 (2): 47–51.

Leyendekkers, J. V., and A. G. Shannon. 2011b. "The Structure of Geometric Number Sequences." *Notes on Number Theory and Discrete Mathematics*. 17 (3): 31–37.

Leyendekkers, J. V., and A. G. Shannon. 2011c. "The Structure of 'Pi'." *Notes on Number Theory and Discrete Mathematics.* 17 (4): 61-68.

Leyendekkers, J. V., and A. G. Shannon. 2012a. "The Modular Ring Z_5." *Notes on Number Theory and Discrete Mathematics.* 18 (2): 2, 28–33.

Leyendekkers. J. V., and A. G. Shannon. 2012b. "Pellian Sequence Relationships among π, e & $\sqrt{2}$." *Notes on Number Theory and Discrete Mathematics.* 18 (2): 58-62.

Leyendekkers, J. V., and A. G. Shannon. 2012c. "Geometric and Pellian Sequences." *Advanced Studies in Contemporary Mathematics.* 22 (4): 507-508.

Leyendekkers, J. V., and A. G. Shannon. 2013a. "On the Golden Ratio." *Advanced Studies in Contemporary Mathematics.* 23 (1): 195-201.

Leyendekkers, J. V., and A. G. Shannon. 2013b. "The Structure of the Fibonacci Numbers in the Modular Ring Z_5." *Notes on Number Theory and Discrete Mathematics*. 19 (1),66–72.

Leyendekkers, J. V., and A. G. Shannon. 2013c. "Fibonacci and Lucas Primes." *Notes on Number Theory and Discrete Mathematics*. 19 (2): 49–59.

Leyendekkers. J. V., and A. G. Shannon. 2013d. "The Pascal-Fibonacci numbers." *Notes on Number Theory and Discrete Mathematics.* 19 (3): 5-11.

Leyendekkers, J. V., and A. G. Shannon. 2014a. The Decimal String of the Golden Ratio. *Notes on Number Theory and Discrete Mathematics.* 20 (1): 27-31.

Leyendekkers, J. V., and A. G. Shannon. 2014b. "Fibonacci Primes." *Notes on Number Theory and Discrete Mathematics*. 20 (2), 6–9.

Leyendekkers, J. V., and A. G. Shannon. 2014c. "Fibonacci Numbers with Prime Subscripts: Digital Sums for Primes versus Composites." *Notes on Number Theory and Discrete Mathematics*, 20 (3): 45–49.

Leyendekkers, J. V., and A. G. Shannon. 2014d. "Fibonacci Numbers as Prime Indicators." *Notes on Number Theory and Discrete Mathematics*. 20(4), 47–52.

Leyendekkers, J. V., and A. G. Shannon. 2015a. "The Golden Ratio Family and Generalized Fibonacci Numbers." *Journal of Advances in Mathematics*. 10(1), 3130–3137.

Leyendekkers, J. V., and A. G. Shannon. 2015b. "The Structural Relationships between Fibonacci and Prime-generated Odd Number Sequences." *Advanced Studies in Contemporary Mathematics*. 25(1), 59–63.

Leyendekkers, J. V., and A. G. Shannon. 2015c. "The Sum of Squares for Primes." *Notes on Number Theory and Discrete Mathematics*. 21(4), 17–21.

Leyendekkers, J. V., and A. G. Shannon. 2015d. "Primes within Generalized Fibonacci Sequences." *Notes on Number Theory and Discrete Mathematics.* 21(4), 56–63.

Leyendekkers, J. V., and A. G. Shannon. 2015e. "The Golden Ratio Family and Generalized Fibonacci Numbers." *Journal of Advances in Mathematics*. 10(1), 3130–3137.

Leyendekkers, J. V., and A. G. Shannon. 2015f. "An Infinite Primality Conjecture for Prime-subscripted Fibonacci Numbers." *Notes on Number Theory and Discrete Mathematics.* 21 (1): 51-55.

Leyendekkers, J. V., and A. G. Shannon. 2015g. "Primes within Generalized Fibonacci Sequences." *Notes on Number Theory and Discrete Mathematics.* 21 (4): 56-63.

Leyendekkers, JV, and AG Shannon. 2016a. "Sequences from $(x^2 \pm y^2)$." *Notes on Number Theory and Discrete Mathematics*. 22 (2): 28-63.

Leyendekkers, J. V., and A. G. Shannon. 2016b. "Some characteristics of the golden ratio family." *Notes on Number Theory and Discrete Mathematics.* 22 (3): 84-89.

Leyendekkers, J. V., and A. G. Shannon.2016c. "Mersenne Numbers." *Bulletin of Mathematics and Statistics Research*. 4 (2): 146-150.

Leyendekkers, J. V., and A. G. Shannon.2016d. "The Goldbach Conjecture." *Journal of Advances in Mathematics*. 11 (9): 5061-5065.

Leyendekkers, J. V., and Shannon, A.G. 2017a. "Primitive Pythagorean triples and generalized Fibonacci sequences." *Notes on Number Theory and Discrete Mathematics.* 23 (1): 54-62.

Leyendekers, J. V., A. G. Shannon and Rebecca A, Hillman, 2017b. "The Brocard-Ramanujan Diophantine equation." *Advanced Studies in Contemporary Mathematics*. 17 (4): 27 (4): 571-575.

Leyendekkers, J. V., A. G. Shannon and J. M. Rybak. 2007. *Pattern Recognition: Modular Rings and Integer Structure.* North Sydney: Raffles KvB Monograph No.9.

Libri, Guillaume. 1838. *Histoire des Sciences Mathématiques en Italie, depuis la Renaissance des Lettres.*[*History of Mathematical Sciences in Italy, since the Renaissance of the Letters*] Paris: Jules Renouard.

Lilley, T. H. 1973. "Raman Spectroscopy of Aqueous Electrolyte Solution." In W*ater: A Comprehensive Treatise, Volume Three,* edited by Felix Franks. New York: Plenum Press, Ch. 6.

Livio, Mario. 2002. *The Golden Ratio*. New York: Golden Books.

Lucas, E. 1969. *The Theory of Simply Periodic Numerical Functions.* (Edited by D. A. Lind; translated by S. Kravitz.) San Jose: The Fibonacci Association.

Lucas, E. 1878. "Théorie des Fonctions Numériques Simplement Périodiques." *American Journal of Mathematics.* 1: 184–240.

Mack, J. M. 1970. "The Continued Fraction Algorithm." *Bulletin of the Australian Mathematical Society.* 3: 413–422.

MacMahon, Percy A. 1915. *Combinatory Analysis, Volume I.* Cambridge: Cambridge University Press.

McCartney, Mark. 2017. The answer is probably e, π or √2. *International Journal of Mathematical Education in Science and Technology.* 48 (5): 768-773.

McClenon, R. B. 1919. "Leonardo of Pisa and his Liber Quadratorum." *American Mathematical Monthly.* 26 (1): 1-8.

McDaniel, Wayne L. 1974. "Perfect Gaussian integers." *Acta Arithmetica.* 25 (2): 137-144.

McInerny, Ralph. 1983. "Beyond the Liberal Arts." *The Seven Liberal Arts in the Middle Ages*, edited by David L. Wagner. Bloomington: Indiana University Press.

McInerney, Stephen. 2017. "Thinking person's guide to what it means to be human": *The Australian.* 8 November, p. 32.

Maritain, Jacques, and Mortimer Jerome Adler. 1940. *Scholasticism and Politics.* New York: The Macmillan Company.

Maritain, Jacques. 1932. *The Degrees of Knowledge*, London: Bles, p. 5.

Maritain, Jacques. 1938. *True Humanism* (Translated by M. R. Adamson). London: Bles.

Matthys, Justin and Richard Wilsom. 2016. *Impact Report: for educators, students, parents and lovers of mathematics.* Melbourne, VIC: Maths Pathway.

Melham, R. S. 2016. "Sums of certain products of Fibonacci and Lucas numbers." *The Fibonacci Quarterly.* 54 (4): 296-303.

Monzingo, M. G. 1980. An Observation Concerning Whitford's 'Binet's Formula Generalized.' In *A Collection of Manuscripts Related to the*

Fibonacci Sequence edited by V. E. Hoggatt Jr and Marjorie Bicknell-Johnson. Santa Clara, CA: The Fibonacci Association, pp. 93-94.

Moore, G. A. 1993. "A Fibonacci polynomial sequence defined by multidimensional continued fractions; and higher order golden ratios." *The Fibonacci Quarterly*. 31(4): 354–364.

Moser, L., and E. L. Whitney. 1961. "Weighted Compositions." *Canadian Mathematical Bulletin.* 4 (1): 39-43.

Nef, John U. 1959. "The University of Chicago and the World (1929-1951)." In *The Image of Man,* South Bend, IN: Notre Dame University Press, pp. 407-33.

Newman, John Henry. 1992. *Conscience, Consensus and the Development of Christian Doctrine.* (Commentary and notes by James Gaffney). New York: Doubleday, Ch. 1, sect. 2, no. 7.

Newman, John Henry. 1996. *The Idea of a University.* (Edited by Frank M. Turner). New Haven and London: Yale University Press.

Newman, M., D. Shanks, and H. C. Williams. 1980/81. "Simple Groups of Square order and an Interesting Sequence of Primes." *Acta Arithmetica.* 2: 129–140.

Nimbran, Amrik Singh. 2010. "On the Derivation of Machin-like Arctangent Identities for Computing Pi (π)." *The Mathematics Student.* 79: 171–186.

Nussbaum, Martha C. 2010. *Not for Profit: Why Democracy Needs the Humanities.* Princeton: Princeton University Press, p. 10.

Omey, E., S. and Van Gulck. 2015. "What are the last digits of...?" *International Journal of Mathematical Education in Science and Technology.* 46 (1):, 147–155.

Panwar, Yashwant K., Bijendra Singh, and V. K. Gupta. 2014. "Generalized Fibonacci Sequences and Its Properties." *Palestine Journal of Mathematics.* 3 (1): 141-147.

Peelle, H. A. 1975. "Euclid, Fibonacci and Pascal – Recursed!" *International Journal of Mathematical Education in Science and Technology*. 6 (4): 395–405.

Philippou, Andreas. 1983. "A note on the Fibonacci sequence of order *K* and the multinomial coefficients." *The Fibonacci Quarterly*. 21 (2): 82-86.

Phillips, George M. 2005. *Mathematics Is Not a Spectator Sport.* New York: Springer.

Pollard, Vicki, Karen Young, Rachael Hains-Wesson. 2017. "Creative Teaching in STEM." *Teaching in Higher Education.* http://dx.doi.org/10.1080/13562517.2017.13794787. pp. 1-16.

Radford, William Cropley. 1964. *A Field for Many Tillings: Research in Education in Australia Today.* Hawthorn: Australian Council for Educational Research.

Ribenboim, Paulo. 2000. *My Numbers, My Friends: Popular Lectures on Number Theory*. Berlin, Springer-Verlag.

Riemann, B. 1859. "Űber die Anzahl der Primzahlen unter einer gegebenen Grösse." ["About the number of primes under a given size."] *Monatsberichte der Königlichen Preussische Akademie des Wissenschaften zu Berlin*, 671-680.

Riesel, H. 1994. *Prime Numbers and Computer Methods for Factorization. 2nd Edition – Progress in Mathematics 126.* Boston: Birkhäuser.

Riordan, J. 1968. "A note on a *q*-extension of ballot numbers." *Journal of Combinatorial Theory*. 4: 191-193.

Roberts, J. 1992. *Lure of the Integers.* Washington, DC: Mathematical Association of America.

Rota, G.-C., P. Doubilet, C. Greene, D. Kahaner, A. Odlyzko, and R. Stanley, R. 1975. *Finite Operator Calculus*. London: Academic Press, Ch. 2.

Sagar, B. 2017. "Extending the beauty of Ramanujan's number." *The Mathematics Student*. 86 (3-4): 97-110.

Sándor, Jóseph. 1984. "On Jordan's arithmetic function." *Mathematics Student.* 52 (1-4): 91-96.

Sándor, Jóseph. 1995. "On a new arithmetic function." *Notes on Number Theory and Discrete Mathematics.* 1 (2): 60-65.

Sarvate, Dinesh J., and Li Zhang. 2014. "Two combinatorial questions." *Bulletin of the Institute of Combinatorics and its Applications*. 70: 45-56.

Sburlati, Giovanni. 2002. "Generalized Fibonacci Sequences and Linear Congruences." *The Fibonacci Quarterly*. 40: 446-452.

Schinzel, A. 1990. "Special Lucas sequences, including the Fibonacci sequence, modulo a prime." In *A Tribute to Paul Erdős* edited by A. Baker, B. Bollabás, A. Hanjel. Cambridge: Cambridge University Press, 1990, pp. 349-357.

Schröder, Bernd S. W. 2002. *Ordered Sets: An Introduction*, Boston, MA: Birkhäuser.

Sellars, Maura. 2017. "Numeracy across the curriculum: A pathway to crtical thinking." *International Journal of Innovation, Creativity and Change*. 3 (2): 75-83.

Shallit, J. O. 1985. "On Infinite Products Associated with Sums of Digits." *Journal of Number Theory*. 21 (2): 128–134.

Shanks, D., J. W. Wrench Jr. 1962. "Calculation of π to 100,000 Decimals." *Mathematics of Computation*. 16: 76–99.

Shannon, A. G. 1968. *Forces and Interaction. (ACER JSSP Volume 4.)* Melbourne, Vic: F. W. Cheshire.

Shannon, A. G. 1974. Some properties of a fundamental recursive sequence of arbitrary order. *The Fibonacci Quarterly*. 12 (4): 327–335.

Shannon, A. G. 1975. "Fibonacci analogs of the classical polynomials." *Mathematics Magazine*. 48 (3): 3–7.

Shannon, A. G. 1977. "Tribonacci Numbers and Pascal's Pyramid." *The Fibonacci Quarterly*. 15 (3): 268, 275.

Shannon, A. G. 1980. "Some Lacunary Recurrence Relations." *The Fibonacci Quarterly*. 18 (1): 73–79.

Shannon, A. G. 1983. "Intersections of second order linear recursive sequences." *The Fibonacci Quarterly*. 21(1), 6–12.

Shannon, A. G. 2004. "Some properties of Fermatian numbers." *Notes on Number Theory and Discrete Mathematics,* 10 (2): 25-33.

Shannon, A. G. 2010. "A Note on Some Diagonal, Row and Partial Column Sums of a Zeckendorf Triangle." *Notes on Number Theory and Discrete Mathematics.* 16 (2):2, 33–36.

Shannon, A. G. 2013. "The Sequences of Horadam, Williams and Philippou as Generalized Lucas Sequences." *Advanced Studies in Contemporary Mathematics.* 23 (3): 551-558.

Shannon, A. G. 2017. "The Fibonacci Sequence and the Jacobian Matrix in Food Web Models." *International Journal of Biosensors & Bioelectronics.* 3(2): 00061. DOI: 10.15406/ijbsbe.2017.03.00061.

Shannon, A. G., P. G. Anderson, and A. F. Horadam. 2006. "Properties of Cordonnier, Perrin and Van der Laan Numbers." *International Journal of Mathematical Education in Science and Technology.* 37 (7): 825–831.

Shannon, A. G., and Leon Bernstein. 1973. "The Jacobi–Perron Algorithm and the Algebra of Recursive Sequences." *Bulletin of the Australian Mathematical Society.* 8(4): 261–277.

Shannon, A. G., C. K. Cook, and R. A. Hillman. 2015. "An Extension of Some Results due to Jarden." *Journal of Advances in Mathematics.* 9 (8): 2937–2945.

Shannon, A. G., and A. F. Horadam. 1973. "Generalized Fibonacci number triples." *American Mathematical Monthly.* 80: 187–190.

Shannon A. G., and A. F. Horadam. 1991. "Generalized staggered sums." *The Fibonacci Quarterly,* 29 (1): 47–51.

Shannon, A. G., and A. F. Horadam. 1994. "Arrowhead curves in a tree of Pythagorean triples." *International Journal of Mathematical Education in Science and Technology.* 25 (2): 255–261.

Shannon, A. G., A. F. Horadam, and S. N. Collings. 1974. "Some Fibonacci Congruences." *The Fibonacci Quarterly.* 12 (4): 351–354.

Shannon, A. G., and J. V. Leyendekkers. 2012. "Pythagorean Fibonacci Patterns." *International Journal of Mathematical Education in Science and Technology.* 43: 554–559.

Shannon, A. G., and J. V. Leyendekkers. 2014. "Extensions to the Zeckendorf triangle." *Notes on Number Theory and Discrete Mathematics.* 20 (5): 31-34.

Shannon, A. G., and J. V. Leyendekkers. 2015. "The Golden Ratio Family and the Binet equation." *Notes on Number Theory and Discrete Mathematics*. 21 (2): 35–42.

Shannon, A. G., and J. V. Leyendekkers. 2016. "Unit fractions and the Erdös-Straus conjecture." *Journal of Advances in Mathematics* 12(6): 6295-6303.

Shannon, A. G., and J. V. Leyendekkers. 2017. "On Legendre's Conjecture." *Notes on Number Theory and Discrete Mathematics*. 23 (2): 117-125.

Shannon, A. G., and Kenny Pruitt. 2017. Functional difference equations. *International Journal of Advances in Mathematics*. 2017 (5): 22-30.

Shannon, A. G., J. C. Turner, and K. T. Atanassov. 1991. "A generalized tableau associated with coloured con-volution trees." *Discrete Mathematics*, 92: 329–340.

Shannon, A. G., and C. K. Wong. 2010. "Some Properties of Generalized Third Order Pell Numbers." *Congressus Numerantium*. 201: 345–354.

Shapiro, L. W., S. Getu, W.-J. Woan, and L. C. Woodson. 1991. "The Riordan Group." *Discrete Applied Mathematics*. 34: 229–239.

Sierpinski, W. 1964. « Sur les Nombres Pentagonaux. » ["On the Pentagon Numbers. "] *Bulletin de la* Société *Royale* des *Sciences de Liège. 33: 513–517.*

Simons, C. S., and M. Wright. 2007. "Fibonacci Imposters." *International Journal of Mathematical Education in Science and Technology*. 38(5): 677–682.

Singh, Bijendra, and Omprakash Sikhwal. 2010. Fibonacci-triple sequences and some fundamental properties. *Tamkang Journal of Mathematics*. 41 (4): 325-333.

Singh, Mamta, Shikha Bhatnagar, and Omprakash Sikhwal. 2012. Multiplicative triple Fibonacci sequences. *Applied Mathematical Sciences*. 6 (52): 2567-2572.

Sloane, N .J. A. 1973+. *The On-Line Encyclopedia of Integer Sequences*; published electronically at http://oeis.org.

Sloane, N. J. A., Simon Plouffe. 1995. *The Encyclopedia of Integer Sequences*. San Diego, CA: Academic Press; and https://oise.org/.

Smale, S. 1998. "Mathematical Problems for the Next Century." *Mathematical Intelligencer,* 20(2): 7-15.

Solzhenitsyn, Alexandr. 1972. *One Word of Truth* (the 1970 Nobel speech on literature; translated by Thomas P. Whitney). New York: Harper and Row.

Somer, L. 2002. "Generalization of a Theorem of Drobot." *The Fibonacci Quarterly*. 40 (4): 435–437.

Southern, R. W. 1970. *Medieval Humanism and Other Studies*, Oxford: Oxford University Press.

Spencer, Adam. 2017. *The Number Games*. Sydney: Xoum Publishing.

Stakhov, Alexy, Samuil Aranson, and Scott Olsen. 2017. *The "Golden" Non-Euclidean Geometry*. Singapore: World Scientific, Chs, 1, 5.

St. Augustine 1997. *Confessions.* (Translated by Maria Boulding, *osb*; edited by John E. Rotelle, *osa*). Hyde Park, NY: New City Press, III, 4, 7.

Stein, S. K. 1962. "The intersection of Fibonacci sequences." *Michigan Mathematical Journal*. 9: 399–402.

Stein, S. K. 1963. "Finite models of identities." *Proceedings of the American Mathematical Society.* 14: 216–222.

Stephens, M. 2011. "Ensuring instruction changes: Evidence based teaching - How can Lesson Study inform Coaching, Instructional Rounds and Learning Walks." *Journal of Science and Mathematics Education in Southeast Asia*, 34 (1): 111-133.

Stillwell, John. 2018. *Reverse Mathematics: Proofs from the Inside Out.* Princeton, NJ : Princeton University Press.

Stormer, C. 1899. «Solution complet en nombres entiers de l'equation $m.\arctan g[1/x] + n.\arctan g[1/y] = k[\pi/4]$» ["Complete integer solution of the equation m.arctan g [1 / x] + n.arctan g [1 / y] = k [π / 4]"] *Bulletin de la Société Mathématique de France.* 27: 160–170.

Subba Rao, K. 1954. "Some Properties of Fibonacci Numbers – I." *Bulletin of the Calcutta Mathematical Society*. 46, 253–257.

Subba Rao, K. 1959. "Some properties of Fibonacci numbers – II." *The Mathematics Student*. 27 (1): 19–23.

Sylvester, James Joseph. 1912. *The Collected Papers. Volume IV (1882-1897)*. Cambridge: Cambridge University Press, Chs. 1, 4.

Terr, D. 1996. "On the Sums of Digits of Fibonacci Numbers." The *Fibonacci Quarterly.* 34(4): 349–355.

Todd, John. 1949. "A Problem on Arc Tangent Relations." *The American Mathematical Monthly.* 56: 517–528.

Townsend, Greg. 2015. "Mathematics in the Liberal Arts Tradition." *Instaurare – The Christendom College Quarterly Magazine*. 23 (2): 24-25.

Trigg, Charlers W. 1980. "Reiterative routines applied to 1979". In *A Collection of Manuscripts related to the Fibonacci Sequence* edited by Verner E. Hoggatt Jr, and Marjorie Bicknell-Johnson. Santa Clara CA: The Fibonacci Association, pp. 90-92.

Wang, Tsung Juang, Kuo Hung Huang. 2017. "Pedagogy, philosophy and the question of creativity." *Teaching in Higher Education.* http://dx.doi.org/10.1080/13562517.2017.1379479. pp. 1-13.

Vajda, S. 1989. *Fibonacci and Lucas Numbers, and the Golden Section: Theory and Applications*. Chichester: Ellis Horwood.

Wagner, David L. 1983. "The Seven Liberal Arts and Classical Scholarship." In *The Seven Liberal Arts in the Middle Ages* edited by David L. Warner. Bloomington, IN: Indiana University Press, p. 24.

Walsh, David J. 1985. "Restoring the Lost Center of Education." *Cooperation in Education*. 47/48: 99.

Ward, Morgan. 1931. "The Algebra of Recurring Series." *Annals of Mathematics*. 32 (1): 1–9.

Watkins, J. J. 2014. *Number Theory: A Historical Approach.* Princeton and Oxford: Princeton University Press.

Wetherfield, Michael Roby. 1996. "The Enhancement of Machin's Formula by Todd's Process." *The Mathematical Gazette.* 80: 333–344.

White, John Joiner. 1998. *A Humean Critique of David Hume's Theory of Knowledge.* (Edited by John A Gueguen). Lanham, New York, Oxford: University Press of America, pp.165-173.

Whitehead, Alfred North. 1925. “The Origins of Modern Science.” In *Science and the Modern World edited by* Alfred I Tauber. New York: Macmillan. pp. 53-69.

Whitford, A. K. 1977. “Binet’s Formula Generalized.” *The Fibonacci Quarterly*. 15(1): 21, 24, 29.

Williams, H. C. 1998. *Édouard Lucas and Primality Testing – Canadian Mathematical Society Monographs 22*. New York: Wiley.

Wolfe, Christopher. 1985. “A Liberal Arts Education Can Be Realized Today.” *Cooperation in Education.* 47/48:105.

Wong, C. K., T. W. Maddocks. 1975. “A Generalized Pascal’s Triangle.” *The Fibonacci Quarterly*. 13 (2): 134–136.

Wooldridge, Kent. 1979. Values taken many times by Euler’s phi-function. *Proceedings of the American Mathematical Society*. 76 (2): 229-234.

Zahn, Maurice. 2017. “A deduction of the Golden Spiral equation via powers of the Golden Ratio.” *International Journal of Mathematical Education in Science and Technology*. 48 (6): 963-971.

Zaremba, S. K. 1970. “A Remarkable Lattice Generated by Fibonacci Numbers.” *The Fibonacci Quarterly*. 8(2: 185–198.

Zeitlin, D. 1965. “Generating Functions for Products of Recurrence Sequences.” *Transactions of the American Mathematical Society*. 116: 300-315.

Brief Biographies of Authors

Tony Shannon is an Honorary Fellow of Warrane College at the University of New South Wales. He also does consulting work at various Australian higher education providers including the Australian Institute of Music, Elite Education Institute, and Kenvale College of Hospitality. He was formerly Deputy Chancellor of the University of Notre Dame Australia where he continues as a Trustee and a Governor. He is an Emeritus Professor of Applied Mathematics of University of Technology Sydney where he was also Foundation Dean of the Graduate Research School. He holds the degrees of *PhD*, *EdD* and *DSc* with research interests in education and epidemiology as well as mathematics. He has had sabbatical leaves in a number of countries and continues to collaborate with colleagues from around the world. He has recreational interests in music, theatre, thoroughbred racing, rugby and cricket.

Jean Leyendekkers was awarded the higher doctorate, *Doctor of Science* (*DSc*), degree by the University of Sydney for her research on Solution Theory. Since retiring from the Faculty of Science there, Jean has written papers on Number Theory and now has eighty seven published. A few of the earlier papers were written with Janet Rybak, but most have been co-authored with Professor Tony Shannon. The emphasis has been on Integer Structure influence on problems in Number Theory. Jean has also been very active for more than thirty years in community work where she lives on urban planning, and in particular on the regeneration of bushland. Jean enjoys classical music, mysteries and loves cats, dogs, possums and all other animals and birds.

INDEX

I

K

L

M

P

R

S

T

Z